T0331205

Culture by Design

Culture by Design is about shifting focus from solely organisational outcomes and performance, towards organisational culture and wellbeing. It bridges the gap between two key organisational goals: (a) the drive for improving performance, outcomes and staff retention, and (b) strategies to encourage employee wellbeing, motivation and engagement within the workplace. For too long, organisations have focussed on each of these goals individually, with improvements in one area often coming at the expense of the other. This book demonstrates that this does not need to be the case, that what is required is a shift in perspective towards a culture-focussed approach where improved outcomes, performance and engagement are the added bonuses of a happy, connected staff team. You may be familiar with the phrase "Happy bees work harder"; this book demonstrates the fundamental truth in that statement and illustrates that "What is good for the bees is good for the hive".

Through practical strategies and real-world examples, this book reveals that the application of evidence-led, self-directed and cost-effective strategies can support any organisation to cultivate the culture they need to encourage the outcomes they want. This book offers a synthesis of theory and practice from organisational and social psychology, neuroscience and systems dynamics, alongside examples of practical tools you can start using today, to offer a roadmap to cultivating a workplace culture that supports the wellbeing and performance of the organisation as a whole.

Whether you are an HR director, People Manager, C-Suite Team member or Wellbeing and Culture Lead, this book is relevant to Leaders in organisations of any size. If you are interested in what works when it comes to improving staff wellbeing, how to go about the process of culture change or who makes the tea and why it matters, then this book is for you.

Hugo Metcalfe is a Psychologist, Psychotherapist, Culture Change Specialist and Co-Founder of the wellbeing consultancy The Happy Mind Tribe. He has designed and delivered evidence-based training programmes and workshops to meet unique organisational needs around the globe. In his pursuit of greater awareness, he felt it prudent to study every type of Psychology going, from Forensic, to Clinical and Occupational Psychology and, more recently, Psychotherapy. He subsequently combined this with over 20 years' experience in the field of mental health and wellbeing, in the UK's National Health Service and private sector. This work has provided him evidence-based insight into 'what works' when it comes to intervention and support at an individual and organisational level.

Over the years Hugo has delivered training, consultancy and research projects to a diverse range of organisations, including Disney, Viacom, Bentley, GSK, the Council of Europe, the UK's Home Office and the British Military. He believes that knowledge is for sharing and he shares his expertise in key areas such as resilience, leadership, stress management, psychological safety, psychological contracts and communication, to create actionable strategies to support organisations to cultivate workplace cultures that work. A multiple world-record holder and expedition guide, Hugo spends his free time putting his own resilience to the test by leading expeditions around the world.

"Hugo Metcalfe's book *Culture by Design* is a must-read for leaders looking to maximise employee well being through lasting culture change. With his extensive experience as a psychologist and well being consultant, Hugo delivers complex concepts in a clear and accessible way. This book provides clear, practical strategies, to create the positive and thriving workplace cultures that are essential for growth."

Major (Retired) Menucha Knebel MSc CMgr,
Lebenk Leadership Consultancy, United Kingdom

"This book has the rare skill of taking deep insights from academia, and making them useful to the everyday leader. Easy to understand, but with more than enough to satisfy the most seasoned practitioners. Use it to maximise you, your team and your business goals."

Henry C. Blanchard, *Entrepreneur-in-Residence,*
University of Staffordshire and Santander X,
United Kingdom

Culture by Design

Practical Strategies for Wellbeing,
Engagement and Growth

Hugo Metcalfe

Routledge
Taylor & Francis Group

LONDON AND NEW YORK

Designed cover image: Orbon Alija/Getty Images

First published 2025
by Routledge
4 Park Square, Milton Park, Abingdon, Oxon OX14 4RN

and by Routledge
605 Third Avenue, New York, NY 10158

Routledge is an imprint of the Taylor & Francis Group, an informa business

© 2025 Hugo Metcalfe

British Library Cataloguing-in-Publication Data
A catalogue record for this book is available from the British Library

Library of Congress Cataloging-in-Publication Data
Names: Metcalfe, Hugo, author.
Title: Culture by design: practical strategies for wellbeing,
engagement and growth / Hugo Metcalfe.
Description: Abingdon, Oxon; New York, NY: Routledge, 2025. |
Includes bibliographical references and index.
Identifiers: LCCN 2024036252 (print) | LCCN 2024036253 (ebook) |
ISBN 9781032526270 (hardback) | ISBN 9781032526225 (paperback) |
ISBN 9781003407577 (ebook)
Subjects: LCSH: Corporate culture. | Organizational sociology. |
Employee motivation. | Quality of work life.
Classification: LCC HD58.7 .M4736 2025 (print) | LCC HD58.7 (ebook) |
DDC 302.3/5—dc23/eng/20241003
LC record available at https://lccn.loc.gov/2024036252
LC ebook record available at https://lccn.loc.gov/2024036253

ISBN: 978-1-032-52627-0 (hbk)
ISBN: 978-1-032-52622-5 (pbk)
ISBN: 978-1-003-40757-7 (ebk)

DOI: 10.4324/9781003407577

Typeset in Bembo
by codeMantra

Contents

Acknowledgements

A heartfelt thank you to Guy, Heather and the team at Taylor & Francis for believing in this project and helping me realise that maybe I was onto something after all.

Huge gratitude also to all the organisations I have had the good fortune to work with around the world through The Happy Mind Tribe, without whom I would not have had the opportunity to grow and develop ideas and see the process of conscious culture change first-hand. Thank you also to Jo; teamwork makes the dream work.

A special thanks to my two greatest sources of inspiration: first, my grandmother Maryse, a psychologist, a feminist and someone who was always willing to listen to my ideas from an early age. Second, my oldest friend Menucha, for helping me 'learn to love discipline' in order to successfully get this book from my brain to the page.

Finally, thank you to my partner, Maisie, for her unending support, care, and love. Thank you for tolerating my absence over so many weekends, as I hid away in my office furiously tapping at a keyboard and for listening to me rant with enthusiasm and frustration about the need to start humanising workplaces. You are loved.

Introduction

Before I was a Wellbeing Consultant and a Culture Specialist, I was first and foremost a Psychologist. The psychological perspective is one that permeates much of my life and most of this book. For me, it is a way of seeing the world and the people in it – understanding the individual, with their individual needs and beliefs, as well as the interpersonal connections and interactions between them and the larger interconnected system in which they all exist.

At the start of my career in mental health, I worked in a number of community Mental Health Teams in a range of health-care settings in the UK, including Forensic, Adult, and Children and Adolescent Mental Health Services. I loved the work, but gradually the system that surrounded what I did, and the service we provided, became more and more constrictive. I felt we simply didn't have the resources we needed to do the job. With services becoming more and more oversubscribed, the criteria for access became stricter and stricter; before I eventually left the National Health Service (NHS), it had essentially become "If you haven't attempted suicide in the last week, you're not coming in", and we still had a full service and a full list of referrals. I was frustrated. I felt I had all this knowledge and information but could only give it out to select people who made it through the door.

I was left thinking "How can I make this more sustainable? How can I reach more people with the same amount of effort?". For me that was training. Finding spaces to share more ideas with more people to have more of an impact. Taking academic and clinical experience and combining it with industry knowledge to create practical and accessible strategies to support people's wellbeing. The workplace seemed like the most logical place to do this, sharing wellbeing tools in the space where most of us will spend the majority of our lives.

Off I went, a spring in my step and glint in my eye, designing and delivering workshops on Wellbeing, Stress Management and Resilience to a diverse range of organisations from charities to prestige car companies. I capitalised on my psychology and therapy background to draw from the evidence base to design wellbeing tools that actually worked and had a measurable impact on people's wellbeing. But it didn't stick. I would find myself invited back to the same companies year on year to provide the same training again and again. I'd trained the managers, I'd trained the staff; why wasn't it working? Maybe it was the length of the workshops, so I tried increasing them to half-day and full-day programmes on Resilience. No, it still didn't work. The same problems still existed; staff would experience a temporary boost in wellbeing, and then it would drop back down to where it had been before. I re-designed the programmes, came up with new tools based on the latest research and drew in experts to try and make the programmes more impactful. Still, the same issue.

Things changed when I was invited to conduct a research project within a branch of the UK military, around the construct and trainability of 'Leadership'. They were looking to evaluate their existing leadership development programme and identify what aspects of Leadership could be taught. I was in my element, sifting through the last 50 years of research to fish out what on earth a 'leader' was and what aspects of this skill could be woven into a training programmes. In the process, I spent a lot of time examining the complex system that is the military, its leadership hierarchy, its practices, behaviours and expectations. For the first time, I was able to get a good look at a complex organisational culture, and it finally made sense.

Plugging in training is all well and good to attempt to address 'issues' that pop up; however, the more I peered into the depths of an organisation's culture, the more I could see where these issues stemmed from. I realised that I had been viewing the employees, with their attitudes, beliefs and behaviours, as separate from the system that surrounded them, the culture that provided a framework to influence these qualities either positively or negatively. I had been thinking too small. I had been thinking like a trainer rather than a psychologist. The same had been true of my time in the NHS; the pervasive culture I experienced was one of bullying, high pressure, finger-pointing and blame. Of course it was, resources were stretched, and staff were burnt out.

I therefore shifted my approach to the identification and development of sustainable workplace wellbeing cultures instead. I realised that the organisations I had been working with already had sustainable cultures; they were just ones that sustained the things they didn't want like stress, burnout and low engagement. I became interested in how organisations could redesign or reorientate their culture towards promoting a healthier system that didn't require constant input from external wellbeing programmes. In essence, I was interested in doing myself out of a job.

This is where I have spent most of my time over the last few years. I co-founded a Wellbeing Consultancy called The Happy Mind Tribe and spend the majority of my time designing and delivering practical, cost-effective and lasting strategies to help organisations begin to shift their culture from the one they had, usually defined as "The way we do things round here", to the one they needed.

The idea to write this book came off the back of a couple of conference talks I delivered a few years ago, one called "Wellbeing as a KPI" and the other "Who makes the tea and why it matters?". The first talk focused on the positive impact of a wellbeing-oriented culture on organisational success, finding the truth in that age old phrase 'Happy bees work harder'. The second explored something called Organisational Citizenship Behaviours (OCBs); this describes the psychological attachment or bond that employees develop with their organisation. It encompasses their loyalty, dedication and willingness to contribute to the organisation's goals and objectives. OCBs form part of the secret psychological glue that binds people to organisations and are a key indicator of employee engagement. The purpose of these talks was to encourage leaders to look beyond the usual metrics of organisational performance and success to the benefits of a wellbeing-oriented culture.

This book serves to further facilitate that shift and bridge the gap between two key organisational goals: (a) the drive for improving motivation, performance, outcomes and staff retention and (b) strategies to support employee wellbeing and engagement within the workplace. Often these are viewed as opposing drives, and organisations find that when they focus strategies and interventions in one of these areas, it is at the expense of the other. I want to demonstrate how this does not need to be the case. That what is required is a shift in perspective and the application of evidence-led, self-directed and cost-effective strategies that support organisations to achieve both of these goals simultaneously.

For too long I feel that organisations have focused on each of these drives individually, missing the value of a holistic, whole-organisation and culture-focused approach. Improved outcomes, performance and engagement are the added bonuses of a happy, connected staff team in a sustainable workplace wellbeing culture. It is true that happy bees work harder, and this book offers a combination of theory and practice, alongside examples of practical tools you can start using today, to cultivate a culture that supports the wellbeing of both the bees and the hive.

The chapters in this book are best read sequentially, with each idea building on the last to give an overview of how we can begin to cultivate healthy organisational cultures from the ground up. However, you are welcome to jump in at the points that feel most relevant to you now. I will consider the individual, social and systemic aspects of an organisation's culture and practical steps to influence each in turn. In addition, I will cover a range of theories and constructs that form the foundation of a sustainable culture, such as resilience, motivation, systemic leadership and psychological contracts. You could write a whole book on each of these topics individually; lots of people have. This book isn't here to make you an expert in these topics. It will instead focus on how you can start to use some of this knowledge practically to make a difference to your organisation's culture. I will lean on my experience as a trainer and consultant to provide some practical individual and group activities to embed the learning at the end of each chapter.

I once heard the phrase "A change in behaviour is far more likely to lead to a change in awareness than a change in awareness leads to a change in behaviour". I can't remember who said it, but it stuck with me because it's true. Doing is better than thinking when it comes to learning. So, if you are genuinely interested in culture change, be prepared to be active and do something about it.

1 Emotional, Social and Relational Intelligence: Navigating Interpersonal Dynamics

Your workplace is not 'like a family'. I've heard this analogy used countless times to describe workplace environment, "We're like a family here...". Notions like this are most commonly expressed by charismatic leaders and usually in an attempt to foster a sense of belonging and mutual support. However, a workplace and a 'family' are fundamentally different things. Whereas a 'perfect' family operates within a framework of unconditional acceptance and long-term emotionally motivated relationships, a workplace is a dynamic ecosystem made up of individuals with a diverse range of backgrounds, beliefs, personalities and motivations.

Unlike the more static roles in a family, the collection of individuals who make up an interrelated workplace 'system' brings with it a range of communication styles, problem-solving approaches and motivations. These workplace dynamics are also fluid and can be influenced by organisational goals and culture. Therefore, in an effort to better understand and navigate this veritable labyrinth of human interactions, we need skills in Emotional, Social and Relational Intelligence (ESRI). Understanding your culture starts with understanding that it's made up of people.

In this chapter we will begin to unravel the essence of ESRI, exploring its relationship to interpersonal and organisational dynamics. At its core lies something I came to understand from my years working within mental health services in the UK, both therapeutically and later as a Workplace Culture expert – "All behaviour is communication, and all communication is an expression of need".

This insight offers a frame to better understand the interconnectedness of human behaviour and communication and the link between our actions, emotions and needs. If we can become more interested in the need and less focused on the behaviour, we can start to appreciate that 'something is always being said'. Let me put that statement into context. Imagine a scenario where a colleague arrives late to a team meeting. You may simply dismiss this behaviour as 'lateness'; however, in doing so, you may fail to recognise it as a form of communication. Perhaps it's a signal that they are overwhelmed by competing priorities? Perhaps they don't think this meeting is important? Perhaps they overslept because their newborn is keeping them up at night? Perhaps

DOI: 10.4324/9781003407577-1

they simply got stuck in traffic? Perhaps it is none of these things. However, if we abandon curiosity and focus solely on the behaviour, we miss what else is being communicated. We miss an opportunity for understanding, for greater empathy and connection.

Developing ESRI empowers individuals to decipher the unspoken language of human interaction, going beyond simply what people say and bringing our awareness to the more subtle communications that occur every day. For years, these skills have been framed as 'Soft Skills'. In the next few pages, I will demonstrate that far from soft they are essential.

Emotional Intelligence: What Does It Look Like?

Example 1: Low Emotional Intelligence

Imagine a team in a busy office environment. They have come together in a team meeting to discuss an upcoming project, the project deadline is looming and tensions are high. Sarah is leading the meeting; she can see that tensions are running high and that there is a growing sense of unease. Feeling under pressure from the upcoming deadline and the need for them to make progress in this meeting, she decides to plough ahead. Adopting her 'I'm in charge' tone, she reminds the group they have a lot to cover and that they need to stay focused and get through the agenda quickly and efficiently. As Sarah powers through the agenda item by item, the team becomes increasingly tense and withdrawn. Sarah feels the need to keep the group 'on task' and doesn't see the value of open discussion. As a result, the team's stress levels increase. As the meeting progresses, tension begins to boil over, and differences in opinion and priorities lead to team members clashing. Sarah barely makes it to the end of the agenda within the time, and the meeting finishes with team members feeling frustrated and demoralised.

Example 2: High Emotional Intelligence

This time when Sarah notices that tensions are high and the team is feeling stressed, she pauses and recognises the importance of identifying and addressing the tension before working through the agenda. Saying what she sees, or in this case feels, she says, "It feels like tensions are quite high at the moment; before we move onto the agenda, I think it's important that we take a moment to address some of the frustrations and concerns people might have about this project". She encourages the team to share their thoughts and feelings, reminding them that their perspectives are important to her. Once the team has had a space to express how they feel and be heard, they begin to relax, the tension lessons and they can productively work through the agenda items. They collaboratively explore solutions to some of the project's challenges and are able to progress with a shared sense of purpose.

In the two scenarios outlined above, I wonder which of them is more common in your workplace?

Emotional Intelligence: What Is It?

I imagine that this is the component of ESRI that you are most familiar with. Perhaps you attended a workshop on it once or there was a module included in some leadership training you attended? Emotional Intelligence is talked about most often in workplace contexts, something we all like to think we are good at and are quick to identify is lacking in others (usually our leaders).

Often referred to as EI or EQ (Emotional Quotient), the concept of Emotional Intelligence was first introduced in the 1990s by psychologists Peter Salovey and John Mayor and later developed and expanded by the likes of Daniel Goleman. It describes our ability to identify, interpret, regulate and utilise our emotions as well as identifying, understanding and responding to the emotions of others. You may have heard it said that 'EI is more important than IQ'. From a glance at the data, a number of studies support the idea that people with high EI tend to have better mental health outcomes, better interpersonal relationships and higher wellbeing. In terms of its impact in the workplace, people with higher EI also seem to be more effective leaders, encouraging greater team performance, and are better at managing stress and navigating conflict (Boyatzis & Goleman, 2007; Salovey & Mayer, 1990).

We can understand the mechanism of how this works by appreciating the impact that reasoning about and reflecting on our emotions can have on problem-solving, conflict resolution and group cooperation. If I understand how I and others are feeling, I can begin to anticipate and predict how communication and behaviours will be interpreted and received. The greater your EI, the more able you are to 'read between the lines' in social interaction and consider the impact your emotional experience may have on your thoughts and actions.

I speak here to the importance of self-awareness in EI. This represents the first, and from my perspective the most important, of the five core capabilities in EI outlined by Goleman.

But before I explore those in more detail, let me put this first one into practice and reflect on your level of EI. If you are interested in formal self-assessment of EI, you can find any number of questionnaires online; for now, I want you to answer the following question.

"How Are You Feeling?"

Notice what came to mind when I asked this question. Was it an 'emotion word'? – 'Happy', 'Angry', 'Sad', 'Joyful' and the like. Was it a physical sensation? – 'Tired', 'Hungry', 'Relaxed', 'My leg hurts'. Perhaps it was one of my favourites, the pseudo-emotion – 'Ok', 'So so', 'Alright', 'Not bad' or the UK's favourite, 'FINE'. If you answered with an emotion word, bravo, you are able to accurately identify how you feel; congratulations, you are in the minority. If you answered with a physical sensation, good try; you've got some level of self-awareness. If you answered with a pseudo-emotion, don't beat yourself up too much; this is how most people respond when asked this question. In fact, you can do an experiment if you like and ask the next three people that you see how they are feeling. I imagine the majority will give you some variation of Fine or Ok.

Now part of the reason for this is habit. "How are you doing?"/"How are you feeling?" has almost become a version of hello for many of us. However, another reason for this is a lack of emotional self-awareness and, in some cases, a general inability to identify and express emotions (a concept known as alexithymia, a very medical sounding word for a non-medical thing). At any given moment of any given day, you feel something, you are having an emotional or physical experience. Yet for many of us, a lack of curiosity and self-awareness, a lack of EI, means we are often not able to identify this experience.

So, what about your EI? I want to ask you to reflect on the following, if you like you can rate yourself from 1 to 10, 1 being terrible and 10 being great:

1. *How good are you at identifying and describing what other people are feeling?*
2. *How curious are you about other people and their experience?*
3. *How good are you at managing your emotions in difficult situations?*
4. *How good are you at letting go of mistakes?*
5. *How good are you at motivating yourself?*
6. *How good are you at managing conflict?*
7. *How well do you deal with change?*
8. *How good are you at changing how you feel?*
9. *How would you rate your level of empathy?*

Reflect on your answer. How did you rate yourself? What about your team? Your leaders? How might they answer? These questions explore the five core capabilities of EI: Self-Awareness, Self-Regulation, Motivation, Empathy and Social Skills.

1. **Self-Awareness:** Confidence in recognising your own emotions, strengths and weaknesses, goals, motivations and values and recognising your impact on the emotions of others.
2. **Self-Regulation:** Self-control and adaptability, including the recognition of your own negative or disruptive emotions and controlling or redirecting emotions towards a more productive purpose.
3. **Motivation:** Drive, commitment, initiative and optimism; those with higher EI are better able to motivate themselves to achieve their goals.

Figure 1.1 Emotional intelligence.

4. **Empathy:** The ability to understand others' feelings, identify with the challenges of another and consider the feelings of others in decision making.
5. **Social Skills:** Skills concerning leadership, conflict management and communication. It draws on the first two elements to manage relationships with and motivate others.

Although these capabilities are influenced to a certain degree by innate characteristics, they also represent aspects of the ESRI mindset we can develop with practice. EI is something you might have naturally, but it is also something you can learn. We will explore how towards the end of this chapter.

Social Intelligence: What Does It Look Like?

Example 1: Low Social Intelligence

Another team, another tight project deadline (sound familiar?). The team is stressed, and interpersonal conflicts have begun to emerge. During a team meeting, two team leads, Max and Alice, disagree over the best approach to solving a technical issue with the project. Max, the senior of the two, fails to notice the escalating tension and disregards Alice's concerns. He dismisses her idea and insists that his solution is the most appropriate one. Alice interprets Max's dismissive behaviour negatively and her team feels ignored, the conflict escalates and the two teams become polarised. With no avenue remaining for effective communication and collaboration, the project stalls, deadlines are missed, productivity drops and team morale plummets.

Example 2: High Social Intelligence

This time Max engages his Social Intelligence mindset and recognises the increasing tension in the room. This time instead of dismissing Alice's concerns, he demonstrates empathy by listening to her ideas and seeking to understand her perspective. He seeks to understand the motivations and concerns that are behind their difference in opinion, helping to build trust and encourage collaboration. With a greater understanding of the rationale behind Alice's idea, Max proposes a collaborative solution that incorporates aspects of both of their ideas. This solution not only resolves the conflict but also serves to strength the connection between the two teams. The team is able to move forward with greater group cohesion and renewed focus achieving their project's goals within the deadline.

Social Intelligence: What Is It?

EI's lesser-known sister, Social Intelligence (SI or occasionally SQ), was introduced by psychologist Edward Thorndike in the 1920s and popularised almost 100 years later by Daniel Goleman. Often lumped in with EI, it is a distinct concept that describes knowing oneself and others, forming meaningful connections, navigating social situations effectively, managing conflict, adapting to social norms and managing complex social change. In a nutshell it describes our awareness of others and our response and adaptation to others in social situations (Goleman, 2006).

In the complex social environments of a workplace, you can see how this would present a useful skill set. For leaders to lead effectively, they need to understand, adapt to and be able to relate to the people they are leading. Indeed, the function of a workplace is to bring a collection of individuals together and attempt to get them to work cohesively and collaboratively towards shared goals!

In keeping with tradition, I'd like you to reflect on the following questions; again you can rate yourself from 1 to 10, 1 being terrible and 10 being great:

1. *How good are you at expressing yourself verbally and making yourself understood?*
2. *How good are you at 'reading' social situations?*
3. *How good are you at adapting to different social situations and knowing how to behave?*
4. *How are you at making a 'good' first impression?*
5. *How good are you are sales?*
6. *How good are you at talking to strangers?*

Figure 1.2 Social intelligence.

7. *How good are you at public speaking or presenting?*
8. *How good are you at finding out what motivates people and what makes them tick?*

High sores on each of these would suggest that you have the necessary social awareness and role-playing skills to allow you to understand and fit into different social situations. Much like EI, your level of SI is influenced by your innate characteristics, upbringing, level of neurodiversity and the like. It is also a mindset and skill set you can develop.

Goleman's research divides SI into four core competencies: Situational Awareness, Situation Response, Empathy and Social Skills.

1. **Situational Awareness:** The ability to understand and assess relevant social and situational cues, read situations, identify issues and comprehend social contexts.
2. **Situational Response:** The ability to adapt to changing social situations and to select effective strategies appropriately.
3. **Cognitive Empathy:** The ability to understand and relate to others and take active interest in them, identifying and responding to changes in their emotional state.
4. **Social Skills:** The ability to speak in a clear and convincing manner, knowing what to say and when and how to say it.

Relational Intelligence: What Does It Look Like?

Imagine you are part of a busy marketing team, and each day is filled with a complex array of deadlines, client meetings and teamwork sessions to design and deliver campaigns. The teams don't always see eye to eye, and today is no different. Conflict arises between Sarah and Michelle over how to approach a project for a particularly demanding client. Sarah is convinced that her strategy is the most cost effective, while Michelle has the oppositive view; it's been a busy week, this client is particularly irritating and neither Sarah nor Michelle is willing to budge. With the increase in tension and lack of agreement, communication breaks down and productivity declines.

Fortunately, another team member, Emily, steps in. Instead of ignoring the issue or taking sides, she invites Sarah and Michelle to sit down and discuss their concerns collectively. She listens to both perspectives and validates their feelings. She supports them to identify common goals and areas of agreement. She facilitates a session where they can work on a collaborative solution together. As a result, they find a solution that meets both of their needs and those of the client. The tension is reduced, and the team moves forward with feeling connected and with clearer communication, turning a potential setback into an opportunity for growth and improved collaboration.

Relational Intelligence: What Is It?

The concept of Relational Intelligence (RI or RQ) has been around for a while but has perhaps been more recently popularised by Psychotherapist Esther Perel, and similarly by Psychologist John Gottman, in their work with couples. At its core is the understanding that in a relationship between two people, there are three 'entities' to consider: me, you and the relationship itself, each of whom has needs, goals and expectations.

In the context of the work environment, RI describes the ability to connect with others and establish effective rapport and mutual trust. It describes how we connect, maintain that connect while engaging in work tasks, establish and maintain boundaries, understand individual needs and relationships to work and navigate conflict. It is the ability to bridge the gap of individuality and facilitate cooperation, cohesion and motivation towards a shared goal. It is not just about being a gregarious extraverted 'people person', but it's about being 'people-smart', understanding the nature and value of relationships and building healthy reciprocal dynamics (Gottman & Silver, 2000; Perel & Perel, 2006).

This final piece or the ESRI puzzle has been gaining popularity in recent years primarily due to our rapidly shifting workplaces. A combination of globalisation, digitisation and post-Covid adaptation has meant that workplaces now often include remote teams, digital employees and on-site staff. How many Teams/Zoom meetings have you had in the last week for example? This increase in digital connectivity has resulted, in many cases, in a decrease in personal connectivity. In my work as a consultant, I recall noticing a shift in organisational needs from 'Resilience' and 'Self-care' pre-pandemic to 'Culture' and 'Connectivity' in the aftermath. For many of us, this was one of the key Covid takeaways, our need for social connections. However, RI goes beyond simply an arising need in our changing workplaces. For many of us, the need to interact with stakeholders from a diverse range of backgrounds within the organisation and externally is an essential part of our working life.

As before, I invite you reflect on your own relationship with RI in the following questions. Rate yourself from 1 to 10, 1 being never and 10 being always:

1. *Do you ask others for feedback for the purpose of enhancing your own growth and career?*
2. *Are you usually fully engaged in all your social interactions rather than being distracted?*
3. *Are you able to carry a conversation when in a group of strangers?*
4. *When you finish a meeting, do you reflect on what you could have said or done better?*
5. *When listening to other people, are you fully engaged in their story?*
6. *Are you described by others as accessible or 'easy to talk to'?*
7. *Do you usually ask people questions to learn more about them?*
8. *When giving a presentation, do you think about adapting it to suit the audience?*
9. *Are you as concerned with how you deliver a message as what you deliver?*
10. *Are you usually able to 'change the energy' in a room when needed?*

Again, this is not a formal assessment, but high scores might indicate that you already have a developed awareness of RI and are using it, consciously or not, in your day-to-day activities. The RI mindset is relational in its essence that so much of it is defined by our interactions with others. The three skills that are most congruent with high RI, and the ones you can personally develop, can perhaps be broken down as follows:

1. **Positive Regard:** This phrase is borrowed directly from the work of psychotherapy and the work of Carl Rogers. Known here as Unconditional Positive Regard, it describes a perceptual position where one shows unconditional support and acceptance or another person, regardless of what they say or do. In the context of RI, it refers to the belief that everyone is essentially a 'good person' trying to do the best they can with

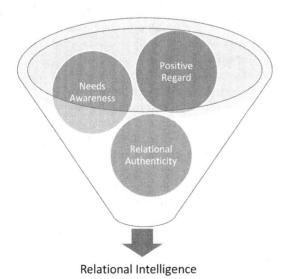

Relational Intelligence

Figure 1.3 Relational intelligence.

the resources that they have and as such they have the right to be treated with respect and care. This perspective values diversity and promotes inclusivity, promoting acceptance over prejudice.

2. **Needs Awareness:** This describes the understanding that all human behaviour stems from their attempts to get their needs met. I remind you here of the statement at the start, the understanding that "All behaviour is communication, and all communication is an expression of need". Someone with high RI is aware of their own needs and curious about the needs or other people, seeking strategies to get them both met in relationship. It supports a reciprocal dynamic of mutual support and growth.

3. **Relational Authenticity:** This refers to an individual's capacity to relate to themselves and others in a genuine, honest and deeper manner – both communicating and inviting honesty in relationships about what's working, what's not and what might need to change. It includes a level of self-awareness and the ability to question what we assume may be true about ourselves or others.

ESRI and Culture

It is time we stop viewing skills like ESRI as 'soft' and instead see them as they are, as essential psychological qualities for successfully navigating the intricate social and relational dynamics of an organisation.

If we can learn to recognise the behaviours of the people that make up our organisation as a form of expression, communicating underlying needs, hopes or struggles, we can look beyond surface behaviours to better understand the individuals who make up our culture. This deeper insight serves to breed deeper connection, empathy and collaboration, fundamentals for a healthy workplace culture. By developing our understanding of ESRI, we can cultivate genuine camaraderie and become more attuned to the psychological state of the organisation.

As well as a useful set of personal skills, greater ESRI is a powerful professional tool to create a fertile environment for positive cultural change. It helps align disparate individuals around common purpose and fosters the openness, trust and goodwill required for teams to thrive. More than just interpersonal skills, these are the competencies that allow an entire organisation to operate as an integrated, emotionally intelligent entity. Any workplace is ultimately an intertwined web of human experiences, motivations and relationships. Developing ESRI provides the insight to begin to align these elements towards an authentically healthy, sustainable organisational culture.

Building Bridges: Practical Strategies for Developing an ESRI Mindset at Work

Developing the core competencies of an ESRI mindset will help foster the skills essential to navigate the complex social dynamics of a workplace ecosystem effectively, enhancing communication and collaboration, improving team dynamics and conflict resolution and enhancing leadership effectiveness. These may sound like lofty goals, but they don't require a psychology degree or years of therapy to accomplish. Below I will outline some simple accessible strategies and activities to begin to develop these competencies. I will focus on the core companies associated with each of the elements of ESRI, delivered through a range of solo and group activities. I invite you to try them out and adapt them to fit your workplace.

Step 1: Develop Self-Awareness

The foundation of ESRI is an understanding of yourself. It includes the ability to identify and understand your own emotions, identify your needs and acknowledge your motivations. You may be exploring this personally already through journaling or mindfulness exercises, or even at work through taking part in personality assessments or 360-degree feedback exercises.

To emphasise the importance of developing these skills, I will borrow the title of John Kabat-Zinn's book on mindfulness meditation – *Wherever you go, there you are* (Kabat-Zinn, 2023). No matter what situation you put yourself in, you will always be you. You take your thoughts, feelings, needs and desires with you, wherever you go. Therefore, to better understand how these things influence your thoughts and behaviours, we need to start noticing them in the first place.

Personal Activity	*Finding Your True Motivation*
Knowing what motivates you is an important aspect of self-awareness that relates directly to you sense of drive, resilience and wellbeing.	
This is a simple exercise that you can do in a few minutes. I want you to ask yourself the question: **"What Motivates You?"**, why do you do what you do, why do you do this job?	
Don't think about it too much, just the first thing that comes to mind. When you think you have an answer, I want you to ask yourself these questions:	
"Why is that important?" & **"What does that give you?"**	
Again, each time you think you have an answer to these questions, ask yourself the questions again. Keep going until you feel you can't go any further or you have reached the 'bottom'.	

Where did you end up? What is behind that initial motivation? What really motivates you?

Let's use a simple example to show you what I mean. Let's say that your first answer to what motivates you was money. I've got bills to pay Hugo. Great. What is money important? What does it give you? Well, Hugo, money means I can pay my bills and I'm not worried about debt. Why is paying your bills important? What does 'no worry about debt' give you? Well, Hugo, that means that me and my family are safe. Why is safety important? What does it give you? Well, Hugo, if I'm safe from worry, then I'm free. Aha, so you are motivated by freedom then, not money.

See what I mean, underneath whatever motivation you think you have found, is a much deeper motivation, something that really drives you forward. If you can take the time to find it, then you have access to an almost inexhaustible supply of motivation. Complete this process for yourself, and then think about members of your team at work. Do you know what motivates them? Do you assume it is promotions, pay raises and job security? Maybe it's not; maybe it's freedom, growth or safety. Ever wondered why the free coffees didn't improve productivity as you hoped? Maybe that's because your team was never motivated by coffee in the first place.

Team Activity	Team Reflection
Objective	Feedback is something many of us seem to have an innate fear of. We associate it with people being 'told off', highlighting our mistakes or being told we need to do better. This is a misconception. Feedback delivered well is something that can fuel connection, mutual appreciation and encourage reflection and self-awareness. In this exercise you are going to get your team to providing constructive feedback to one another.
Duration	1–1.5 hours
Materials Needed	Meeting room or designated space conducive to open discussion

Here's how it works:

Introduction (10 minutes): Introduce the purpose of the workshop. Set expectations of open communication, respect and confidentiality. Outline the agenda and any other ground rules.

Start by bringing your team together. Arrange them in pairs preferably with someone they trust and know well.

Provide Feedback (30 minutes): Each person will then take it in turns to provide feedback to their partner. Remind them that the feedback should focus on specific behaviours, strengths and areas for improvement related to work performance, communication or their social interactions.

Reflect (15 minutes): After receiving feedback, invite everyone to take some time to reflect on the information they've received. Prompt them to consider how their behaviours and actions impact others, what strengths they have and areas where they can grow and develop.

Discuss (20 minutes): Once both partners have received feedback and reflected on it, they come together to discuss their insights. They share their reflections, ask clarifying questions and discuss strategies for leveraging strengths and addressing areas for improvement.

(Continued)

Action Plan (15 minutes): Finally, invite them to develop an action plan based on their reflections and discussions. Encourage them to set specific, measurable goals for enhancing their self-awareness and improving their performance in the workplace.

By engaging in the Feedback Reflection exercise, staff members can gain insight into strengths and areas for development. They can develop their Self-Awareness and SI by learning to see themselves from another's perspective.

Step 2: Foster Empathy

Empathy is a key component of ESRI – the ability to 'feel with people'. We can encourage this skill in team members by engaging in active listening. This practice supports people to identify and acknowledge another's emotions and develop connections by allowing us to 'walk in someone else's shoes'.

Team Activity	Empathy Circles
Objective	This exercise gives staff members with an opportunity to practice empathetic listening in a supportive group setting. Outline the agenda and any ground rules for the session.
Duration	1–1.5 hours
Materials Needed	Meeting room or designated space conducive to open discussion

Introduction (10 minutes): Introduce the purpose of the workshop. Set expectations of open communication, respect and confidentiality. Outline the agenda and any other ground rules.

Gather in Groups: Staff members are divided into small groups of 4–6 people. Do you best to ensure a diverse mix of perspectives and experiences within each group.

Share & Reflect On Personal Stories (60 minutes): Each person in the group takes turns sharing a personal story or experience related to a specific theme or topic of your choosing for the day, such as a challenge they've faced, a success they've achieved or a difficult decision they've had to make. The storyteller speaks uninterrupted for a period of time (3–5 minutes will do), while the rest of the group listens attentively without responding.

After each person shares their story, group members take turns reflecting on what they heard and offering supportive responses. They may express empathy, validate the storyteller's feelings or share how the story made them feel (The listeners should try to avoid sharing similar stories so as not to inadvertently replace the speakers experience with their own). The focus should be on understanding and connecting with the emotions and experiences shared by the speaker.

Debrief and Discuss (20 minutes): Once the activity is complete, the group comes together for a debriefing session. They discuss their experiences, insights gained and challenges encountered during the activity. Group members reflect on how practicing empathy can enhance communication, collaboration and support within the workplace.

By engaging in the Empathy Circles activity, staff members develop their empathy skills by actively listening to others' experiences, sharing in their emotions and offering support and understanding. This fosters a culture of empathy and compassion in the workplace, leading to stronger relationships, increased teamwork and improved overall wellbeing for staff members.

Step 3: Cultivate Active Listening

SI is about understanding and navigating social situations effectively. Fostering a culture of active listening in your organisation supports people to feel heard and valued. The skill of 'listening' is not always one that people spend a lot of time actively developing. It tends to be more of a passive process then an active one. Therefore, providing opportunities to hone these skills can support teams to develop SI, build stronger connections, improve conflict resolution and build trust.

Team Activity	Active Listening Role Play
Objective	This activity gives teams the opportunity to practice Active-Listening Skills.
Duration	30–60 minutes
Materials Needed	Meeting room or designated space conducive to open discussion

Here's how it works:

Introduction (10 minutes): Introduce the purpose of the workshop. Set expectations of open communication, respect and confidentiality. Outline the agenda and any other ground rules.

Put everyone into pairs (a group of three is possible if there are odd numbers). The pairs are assigned one of two roles: Speaker or Listener. The job of the speaker is to speak about a 'difficult or challenging experience' for 5 minutes. The job of the listener is to actively listen to that experience.

Practice Active Listening (20 minutes): As the speaker shares their thoughts, feelings or ideas, the listener practices active listening techniques such as bringing their attention fully to the speaker, maintaining appropriate eye contact and body language to show understanding. The listener should refrain from interrupting or offering advice until the speaker has finished speaking. One the speaker has finished, the listener can practice conscious communication through paraphrasing what has been said to confirm understanding – "It sounds like what you are saying is X… is that right?" After the reflective feedback, they switch roles, and the listener becomes the speaker and vice versa.

Note About Time: Remind people that the whole 5 minutes is theirs, and that if a silence appears that the Listener should leave that silence as long as possible to allow the speaker time to think and to add anything else, they want to add.

(Continued)

Debrief (20 minutes): Once both partners have completed the exercise, encourage them to speak about their experience of the exercise, what aspect they found helpful or unhelpful. Any challenges and how they could apply these skills to a broader work context.

Step 4: Promote Collaboration

RI involves building and maintaining meaningful connections with others. Creating opportunities for cross-functional projects and team-building activities and fostering a culture of appreciation can help encourage teamwork. Working together towards common goals supports organisations in leveraging individual strengths and perspectives of team members in achieving shared success.

Team Activity	*Cross-Functional Problem-Solving*
Objective	This activity encourages staff members from different departments or teams to come together to solve a real-world problem or tackle a specific challenge in the organisation. Engaging in Cross-Functional Problem-Solving exercises like this not only encourages staff to work together to address challenges in the organisation and develop their collaboration skills, but at the end, you might actually have some helpful solutions to solve some problems!
Duration	1.5–2 hours
Materials Needed	Meeting room or designated space conducive to open discussion

Here's how it works:

Introduction (10 minutes): Introduce the purpose of the workshop. Set expectations of open communication, respect, and confidentiality. Outline the agenda and any other ground rules.

Define the Problem (15 minutes): Start by identifying a real-life problem or challenge that your organisation is facing. This could be related to the improvement of processes, customer service or the development of a new project.

Form Cross-Functional Teams (10 minutes): Split people into teams, ensuring a mix of expertise, skills and perspectives in each team. This can be further improved by bring in people from a range of teams across the organisation. Encourage participants to work with colleagues they may not typically collaborate with in their day-to-day roles.

Provide each team with the necessary resources, information and tools to understand the problem and explore potential solutions (this might include data, background information or access to subject matter experts).

(Continued)

Explore Solutions (45 minutes): Give teams time to share ideas and explore potential solutions to the problem. Encourage creative thinking and open-mindedness and remind participants that there are no wrong answers during this phase – "There are no bad ideas in brainstorming". As the team's ideas develop, encourage them to collaborate, drawing on each other's experience to refine their ideas based on collective feedback within the group.

Present Solutions (15 Minutes): Once the groups have developed their potential solutions, invite them to present their ideas to the larger group.

Select and Implement (10 minutes): After all teams have presented their solutions, facilitate a group discussion to select the most viable and impactful solution. Consider factors such as feasibility, cost-effectiveness and alignment with organisational goals. Once a solution is selected, collectively create an action plan for implementation.

Debrief (20 minutes): Finish the exercise with a debrief session. How did they find the exercise, what went well, what could be improved next time and what were the key learning points for future collaboration?

References

Boyatzis, R. E., & Goleman, D. (2007). Emotional intelligence competencies in leadership development. *Journal of Management Development, 26*(5), 437–445.

Goleman, D. (2006). *Social intelligence: The new science of human relationships.* Bantam Books.

Gottman, J. M., & Silver, N. (2000). *The seven principles for making marriage work.* Crown.

Kabat-Zinn, J. (2023). *Wherever you go, there you are: Mindfulness meditation in everyday life.* Hachette UK.

Perel, E., (2006). *Mating in captivity* (p. 272). New York, NY: HarperCollins.

Salovey, P., & Mayer, J. D. (1990). Emotional intelligence. *Imagination, Cognition and Personality, 9*(3), 185–211.

2 Magic Words: The Art of Communication

The art of communication underpins all elements of social interaction. We often demonstrate our emotional, social and relational intelligence through the words we say. The way we communicate shapes culture itself. Productive organisations thrive on open communication, whereas toxic environments are driven by unhealthy communication patterns where people feel unheard. In these environments resentments deepen and cultures rot from within. Effective communication happens when both parties feel that their voice matters, when they feel heard and understood. This is something most organisations understand the importance of, but few make a conscious effort to develop the skills necessary to do it well.

In this chapter I will share a range of communication models and strategies to enhance relationships, provide support and facilitate change. I will draw insights on relevant tools from a range of disciplines such as Occupational Psychology, Psychotherapy, Coaching as well as the invaluable lessons of Marshall Rosenberg's Non-Violent Communication (NVC) and motivational coaching techniques. I will identify some of the core skills that make up effective communication such as open questions, active and reflective listening, showing empathy and encouraging self-efficacy, alongside how and when to use them. In addition, I will explore communication strategies for conflict transformation. The goal? To equip you with the communication tools needed for a healthy culture to thrive.

Magic Words

After stepping from the world of therapy into the world of organisational wellbeing and leadership consultancy, I have often been asked, "Tell me the magic words to say to diffuse conflict, help people change, or to get people to express what they need or how they feel". It seems there is often a belief that if you can say something in just the right way it will foster instant intimacy, connection and trust. This isn't necessarily true. While there are perhaps some magic words when it comes to effective communication, they are only effective when combined with a certain amount of 'magic' awareness.

DOI: 10.4324/9781003407577-2

Firstly, it is important to understand that nearly all conflict and disconnection in an organisation, indeed in life, stems from the same fundamental place, miscommunication. This isn't always an intentional or malicious thing; it describes how two people, groups or teams have attempted communication but missed. Therefore, if we can implement tools to miss less and foster effective communication, we can create a culture of clear communication, improve connection and understanding.

Non-Violent Communication

It sounds a little dramatic, no? The idea is that we should be ensuring that we do not communicate violently. The word 'violence' itself often conjures up distressing images and leaves us feeling uncomfortable. How could my communication, or any communication for that matter, be considered violent? However, if we agree that violence describes any act or behaviour that causes harm to another, we can appreciate how some forms of communication may be considered violent.

This is the central premise behind psychologist and mediator Marshall Rosenberg's NVC construct. He developed the concept in the 1960s influenced by the teachings of Mahatma Gandhi and Martin Luther King Jr. NVC is, in essence, a communication framework designed to encourage authentic and compassionate communication to enhance relationships and resolve conflict between individuals, groups and society. The NVC framework has since been applied in numerous settings including schools, prisons, workplaces and, in the realm of international conflict, resolution (Rosenberg, 2015).

It is based on the notion that all human beings share common needs and that conflict arises when these needs are not met. NVC emphasises the importance of expressing one's needs honestly and empathetically while also being curious about the needs and feelings of others. The framework is divided into four components or processes:

1. **Observations**: This involves describing what we see or hear in a situation without adding judgements or interpretations. It's about stating the facts objectively, without bias or storytelling. For example, "When I observed X happening" rather than "When you were being mean earlier".
2. **Feelings**: This component involves identifying and expressing our emotions in response to the observations we've made. It's about being aware of our feelings and being able to articulate them clearly and honestly. Your self-awareness skills will come in handy here.
3. **Needs**: Here, we identify the underlying needs or values that are behind our emotional experience. It involves appreciating the fundamental human needs that are universal to all people, such as the need for safety, connection, autonomy or respect. These needs should be about us rather than making demands on the other. For example, "I need to feel respected" rather than "I need you to stop being mean".

4. **Requests**: This involves making clear, actionable requests based on our observations, feelings and needs. It's about communicating what we would like to see happen or what actions we would like others to take to meet our needs. Again, it is important that these are requests rather than demands.

NVC: What Does It Look Like?

Let's imagine a workplace scenario. You are in a team meeting with a project deadline looming. You notice that some team members are consistently arriving late to meetings, leading to delays and casing frustration within the group. Instead of reacting with frustration or making assumptions about their intentions or engaging in dismissive or 'violent' communication such as "Why are you late?" or "You must arrive on time", you decide to use the NVC framework to address the issue compassionately.

Observations: You start by objectively stating the observation without judgement. You might say, "I've noticed that in the past few meetings, some team members have arrived late by 10–15 minutes".

Feelings: Next, you express how this behaviour makes you feel. You might say, "I feel frustrated and concerned because it seems like our meetings are starting late, and it's affecting our productivity and team morale".

Needs: Then, you identify the underlying needs or values that are important to you in this situation. You might say, "I value punctuality and efficiency in our meetings because it's important for us to respect each other's time and work collaboratively to meet our deadlines".

Requests: Finally, you make a clear, actionable request based on your observations, feelings and needs. You might say, "I would like to request that everyone do their best to arrive on time to our meetings so that we can start promptly and make the most of our time together. If anyone anticipates being late, can they please let the team know in advance?"

By using NVC in this situation, you address the issue directly and respectfully, focusing on the specific behaviour (late arrivals), expressing your feelings and needs and making a clear request for action. This approach promotes understanding, accountability and collaboration among team members, creating an environment for improved communication and productivity. When used effectively, NVC can offer powerful framework for building bridges and fostering connection in the workplace as well as your personal life.

Motivational Coaching: Communicating To Motivate

Communicating to motivate others to change is an important aspect of communication within organisations, particularly for leaders, many of whom may have likely developed their own brand of motivational communication or 'pep talks'. Do the strategies they have developed actually work? Do staff feel motivated or just more pressured? Motivational communication is a skill, and we can learn to do it better. One strategy to do this is a coaching technique called Motivational Interviewing (MI).

MI is an evidence-based strategy originally developed in therapeutic settings for facilitating behaviour change and inspiring clients to action. It has now transitioned into the workplace in the realms of coaching and leadership. The transition was relatively seamless as the framework supports people to overcome obstacles, identify strength, achieve goals and empower then towards growth and development – skills equally beneficial at work as they are in the therapy room. Developed by William Miller and Stephen Rollnick in their work with addiction, MI is not about pushing people to change; it's about guiding them to find their own motivation through and asking the right questions, listening deeply and having them identify their own potential and barriers to success. It offers a simple and effective collaborative communication tool for supporting change valuing individual autonomy. At its heart is a 'guiding' approach to communication that straddles the line between following (active listening) and leading (giving support and information). MI is most beneficial when people are ambivalent or unsure about change, lack confidence in their ability to change, are unsure whether they want to change or not and are unclear about the benefits and disadvantages of change (Miller & Rollnick, 2012).

MI is a broad concept with a range of tools, techniques and strategies that take time and practice to master. I can't make you an expert in one chapter; however, I can outline the core principles alongside some 'magic words' and communication strategies you can use to support people to change.

MI Core Principles

Expressing Empathy: This skill describes our ability to feel with people, to seek to understand and validate another person's experience without judgement. MI seeks to understand where the person is currently rather than telling them where we want them to be.

Workplace Example: *You are speaking with a team member who is feeling overwhelmed by their workload and struggles to balance work performance and wellbeing.*

Empathy Magic Words:

1. *"It sounds like you're going through a challenging time…"*
2. *"That sounds really tough…"*
3. *"It must be hard to…"*
4. *"You're doing the best you can…"*
5. *"I appreciate you sharing this with me…"*

Developing Discrepancy: In MI it is helpful to support people to identify discrepancies between their desired goals or stated values and their current behaviour, highlighting the 'gap' and the need for change to overcome it.

Workplace Example: You support a team member to explore the misalignment between their career aspirations and their current work habits.

Discrepancy Magic Words:

1. *"You mentioned that you want to [goal], but…"*
2. *"It seems like there's a difference between what you want and what's happening…"*
3. *"Let's explore the gap between where you are now and where you want to be…"*
4. *"What do you think might be getting in the way of you achieving [goal]?"*
5. *"What changes do you think you need to make to bridge the gap between where you are and where you want to be?"*

Rolling with Resistance: Most people are resistant to change; otherwise, they would have done it already. Therefore, instead of confronting or opposing their resistance to change in MI, we explore the resistance or ambivalence without blaming or escalating to conflict.

Workplace Example: *You speak with a staff member who is resistant to seeking help for their performance they are unsure about their ability.*

Resistance Magic Words:

1. *"It's okay to feel unsure or resistant…"*
2. *"Let's explore what's holding you back…"*
3. *"How can I support you in overcoming these obstacles?"*
4. *"What has helped you overcome challenges in the past?"*
5. *"What's one thing you could try differently?"*

Supporting Self-Efficacy: In MI you emphasise and encourage the individual's belief in their ability to change, enhancing their self-belief and confidence.

Workplace Example: *You speak with a staff member who lacks confidence in their ability to succeed in their career.*

Self-Efficacy Magic Words:

1. *"What strengths or skills do you have that can help you in this situation?"*
2. *"What past successes can you draw upon to help you now?"*
3. *"You have the skills and resources to make this change…"*
4. *"Remember when you [past success]? That's evidence of your ability to…"*
5. *"What's one thing you can do today to move closer to your goal?"*

MI Communication Strategies

The four core principles of MI are expressed through four key communication strategies which facilitate the necessary dialogue to improve awareness, uncover goals and strengths and motivate people towards change. I have outlined these alongside a workplace example and some magic words and phrases to help you practise them:

Open Questions: In order to draw out the individual's experiences, perspectives and ideas about the current situation, we can guide the individual to reflect on the possibility of change and what it might mean. Open questions can help someone explore their thoughts and feelings and express themselves freely.

Workplace Example: You are meeting with a staff member who is struggling with time management and feels overwhelmed by their workload.

What Could You Ask?

1. *"What specific challenges do you face when it comes to managing your time effectively?"*
2. *"How do you decide which tasks to tackle first when you have a lot on your plate?"*
3. *"What would you say are the biggest obstacles preventing you from managing your time more effectively?"*
4. *"In an ideal scenario, how would you like to structure your day to better manage your workload?"*
5. *"What support or resources do you feel you need to help you better manage your time?"*

Affirmation: Re-affirming the individual's strengths, positive attributes, past efforts and successes in order to boost their self-esteem and their confidence in their ability to change.

- **Reflections:** Using the skill of active listening to seek to understand the individual's experience, repeating, paraphrasing and demonstrating empathy and understanding of what the person is communicating.
- **Summarising:** Using the active listening skill or summarising what the individual has said to confirm our understanding and help people gain insight and clarity.

Workplace Example: You are speaking with a staff member who is working hard to improve their performance after a drop in performance.

What Could You Say?

1. *"It's clear that you're putting in a lot of effort and commitment to turn things around".*
2. *"Your willingness to seek support and work hard to improve speaks volumes about your character".*
3. *"Your efforts to improve your performance demonstrates your commitment to your role and future success".*
4. *"I appreciate your perseverance and resilience in the face of past challenges".*
5. *"You've made significant progress since the last time we spoke, and that's something to be proud of".*

Reflective Listening: Paraphrasing, summarising or reflecting back what the person has said, demonstrating understanding and empathy and encouraging further exploration.

Workplace Example: You are discussing career aspirations with a staff member who is feeling uncertain about their future.

What Might You Say?

1. *"It seems like you're facing a lot of uncertainty about what comes next in your career journey".*
2. *"You're expressing some hesitancy about making decisions regarding your future career".*
3. *"I understand that you're feeling unsure about which direction to take with your career".*
4. *"It seems like you're in a bit of a dilemma when it comes to mapping out your career path".*
5. *"It sounds like you're seeking clarity and direction in your career decision-making process".*

Summaries: Summaries involve condensing and restating key points from the conversation, helping individuals to gain clarity and insight and reinforce understanding.

Workplace Example: You are discussing a staff member's goals for the upcoming quarter, including professional and personal aspirations.

What Might a Summary Sound Like?

1. *"So, if I understand correctly, your main professional goal for the upcoming quarter is…"*
2. *"If I'm hearing you correctly, your priorities for the upcoming quarter include…"*
3. *"From our conversation, it appears that your primary objectives for the upcoming quarter are…"*
4. *"Let me make sure I've captured everything correctly: you're planning to work towards [X goal] and [Y goal] in the upcoming quarter".*
5. *"To sum up, your goals for the upcoming quarter revolve around [X goal] and [Y goal], with a focus on [any additional goals mentioned]".*

Motivational Interviewing: What Does It Look Like?

Now let's bring this into a workplace context. Imagine you're a manager overseeing a team of sales representatives. One of your team members, Emily, has been struggling to meet her sales targets and seems demotivated. Instead going down the route of disciplinary action or telling her what she needs to improve, you decide to use MI technique to support self-motivated change.

Expressing Empathy: You start by scheduling a one-on-one meeting with Emily. At the start of the meeting, you begin by engaging in empathy and sharing what you have observed and expressing genuine concern for her wellbeing. Using empathy and active listening, you create a support environment where Emily feels comfortable sharing her experience rather than judged. You might say, "It sounds like you have a lot going on at the moment" and "Thank you for sharing this with me".

Developing Discrepancy: As Emily shares her experiences, you support her to identify her goals, what she would like to be different and what might be getting in the way of performing as she would like to be. You might ask questions like, "What would success look like for you in this role?", "What do you think is getting in the way of you being where you want to be?" and "What would need to change for you to be able to perform as you would like to?".

Rolling with Resistance: Emily is worried that she might not be able to improve and perform as she would like to. Instead of opposing her resistance, you explore it with her. You might say things like, "What do you think is holding you back?", "Is there one thing you could change today to start to move towards your goal?" and "What do you need form me to help you overcome these obstacles?".

Supporting Self-Efficacy: In order to support Emily to embed her ideas and plans for change, you work to build her self-esteem and confidence so that she can put her plans into action. You might say things like, "What strength of skills do you have that can help you achieved the goals your outlined?", "I am impressed by your ability to identify what needs to change and believe you have the skill to do it" and "I remember how you performed last quarter and the one before that, I know you have the ability to do this".

Summarise: At the end of the meeting, you draw the key points together and summarise what you have talked about. Reaffirming Emily's commitment to improving her performance and her plan to achieve it, express your confidence in her ability to succeed in reaching her goal. You might say, "Ok, so it sounds like you have identified X & Y as things that are impacting your performance at the moment and you plan to use Z to help you overcome them"., and "I appreciate your honesty and openness today. I'm confident that with your determination and the support you have asked for from me, you can overcome these challenges".

You can see that using a technique like MI can help create a supportive and empowering environment for Emily to explore her current experience, what is getting in the way, her motivations for change and what can help her succeed. Instead of feeling criticised, told off or demoralised, Emily feels heard, valued and motivated to start to take self-directed action towards change. The MI technique values autonomy and self-efficacy, inspiring people to change rather than telling them to.

Transforming Conflict: Communication for Connection and Collaboration

Communication doesn't always go perfectly. As I mentioned at the start of this chapter, sometimes we can 'miss' when attempting to communicate and that can result in conflict. Conflict is an inevitable aspect of workplace dynamics and usually the direct consequence of miscommunication. Workplace conflict can arise from differences in perspectives, goals, access to resources and priorities (De Dreu & Gelfand, 2008). It is often seen as an inevitable consequence of sticking a seemingly random assortment of humans together and telling them to work collaboratively towards a shared goal, a.k.a. a workplace. However, rather than viewing conflict as destructive or unproductive, it is far more helpful to shift our perspective to consider its potential for growth and transformation. In order to make that shift successful, it is important that we learn some tools to support effective communication around conflict and strategies to encourage collaborative problem-solving (Deutsch, 2006; Fisher et al., 2011).

Why 'Conflict Transformation' over traditional 'Conflict Management' or 'Resolution'? Conflict resolution methods tend to focus on compromise and avoidance. Wait, compromise is what we are seeking? No. Stop compromising!

In any conflict situations, you are left with three potential outcomes:

* **Capitulation**: Where one person gives in to the demands or another, a win-lose situation. One person is usually left feeling vindicated and the other unhappy.
* **Compromise**: This is everyone's favourite strategy, the one you were probably taught at school, where the parties agree on a compromise situation where everyone gets a little of what they want perhaps. We need to stop doing this. This is an example of a lose-lose situation. Nobody gets what they actually wanted and both are left feeling slightly disappointed. Eventually resentment will build, and we will return to conflict.
* **Collaboration**: This is the true win-win scenario where both parties find a way to identify and address the underlying issues and get both of their needs met in a transformative solution. Both leave happy.

While compromise may offer a temporary solution to conflicts and is often a quick-fix strategy, collaboration offers a more sustainable approach where we recognise conflict as an opportunity for growth and innovation. By pooling resources and expertise in this way, team members can generate innovative solutions that address the underlying causes of conflict and promote long-term success. While compromise builds resentment, collaboration builds trust and strengthens relationship through greater understanding. The sense of shared ownership from a collaborative solution means that we take personal responsibility for the change and feel connected and invested in it, meaning we are more likely to stick it out.

Sounds good on paper; what about in practice?

Imagine a scenario where a marketing team is divided over a new advertising campaign strategy. Some team members advocate for a traditional approach, while others propose a more innovative and unconventional strategy. Rather than resorting to compromise by settling for a disappointing middle ground, the team decides to seek a collaborative solution that meets everyone's needs. Through a combination of open dialogue, idea sharing and active listening, the team explores a range of different perspectives, developing a hybrid campaign that satisfies their needs and meets their marketing goals. The consequence is a greater sense of connection, people feeling heard and valued and an increased feeling of commitment to the project as it is now shared by the whole team.

Conflict Communication Strategies

So how do we do it? What are the magic works for conflict transformation? Much like our other communication strategies in this chapter, they build on your Emotional, Social and Relational Intelligence skills to communicate empathy, understanding and clarity and practise active listening and the identification and expression of needs. The construct of NVC lends itself extremely well to conflict de-escalation and transformation. Let me break down some of the other helpful approaches and I've some examples;

'I' Statements: 'I' statements are a powerful way to express your thoughts, feelings and needs without blaming or accusing the other person. They start with "I feel" or "I need", followed by a specific description of your emotions or needs. 'I' statements encourage personal responsibility and help prevent defensiveness or escalation. For instance, instead of saying, "You always ignore my suggestion!" you could say, "I feel discouraged when my suggestions aren't acknowledged". A classic example in the workplace is "We are concerned about…", who is we? The whole team, the entire organisations, everyone I have ever met. "I am concerned about…" is far more likely to build connection rather than frustration or fear.

Empathetic Communication: Using your skill of empathy is essential in conflict transformation, especially when we are feeling frustrated ourselves, seeking to understand the needs of the other while validating their experience. It requires genuine curiosity and a willingness to 'view things from their perspective', even if you disagree. By acknowledging their feelings and experiences, empathetic communication builds trust and rapport, the foundation for constructive conflict transformation. You could say, "It sounds like you're feeling overwhelmed by the workload. How can I support you?", "It sounds like you are feeling annoyed by this situation" or "I can see this is something you care about a lot".

T.E.D Language: This stands for Tell, Explain and Describe. This strategy was given to me by a colleague who worked for a suicide prevention charity. She spoke about using it to help draw out more information from an individual to better understand their experience and perspective. You might ask, "Can you tell me a little more about that?", "Can you explain what you mean by that?" and "Can you describe that to me in a little more detail?". These communication strategies can help draw out the necessary understanding of the other person's sense making that can aim us in finding common ground on our path to collaboration.

Identifying Needs: If we appreciate that conflict usually arises from an experience of unmet needs, seeking to understand individuals needs can help us in finding a collaborative resolution where all our needs are met. Remember we can't do that if we don't know what the other person's needs are in the first place. How do you find out what someone's needs are? Ask them, "What do you need right now?", "What do you need from me?", "What do you need from this conversation" or my personal favourite, "What do you need me to understand?". The last one is one of my go-to strategies for diffusing an argument where someone has become frustrated or even aggressive. Clearly, they need you to understand something; otherwise, they would have become so frustrated so why not ask them what that is. Expressing needs allows individuals to articulate their desires, priorities and concerns while also enabling them to better understand the needs of others. Ultimately, this leads to a more empathetic, collaborative and sustainable conflict transformation process.

Collaborative Problem-Solving: If we are seeking collaboration as a conflict transformation outcome, we need to work together with the other person to find mutually beneficial solution. This requires the willingness to explore and share ideas, treat differing perspectives with curiosity and respect and enter with the mindset of finding a solution together. This approach encourages shared responsibility and ownership over the outcome, leading to more sustainable solutions. For example, you could say, "Let's explore some options together to address this issue and find a solution that works for both of us", "I think we are both trying to achieve the same outcome and I'd like to find a way that meets both of our needs" or "What might we need to find a collaborative solution to this problem?".

Communication and Culture

Developing the art of healthy communication is one of the key practices in cultivating positive cultural change. When people feel truly heard and understood, transformation can take root. But this requires practical skills in how we choose to engage socially within the organisation and convey our thoughts and feelings to others. With non-violent communication techniques, we can avoid conflict by learning to make observations without judgement and to voice feelings and needs with empathy, enhancing connection and understanding. You can use MI techniques to bypass confrontation in favour of evoking people's own motivations for positive change, drawing out their hopes and values to build investment in the process.

Now, of course, some conflict is inevitable in any workplace. But approaching it with a lens of transformation allows us to explore different perspectives, validate all viewpoints and seek common ground and collaboration for mutual gains. Used skilfully, these modes of communicating plant the seeds for cultural renovation. They nurture insight, goodwill and collaborative solutions. Real culture change starts by transforming how we engage with one another. Healthy communication ensures all voices get heard, reduces animosity and inspires people to communicate with clarity.

From Magic Words to Practical Magic: Practical Strategies for Improving Communication and Conflict Transformation at Work

In our personal and professional lives alike, the way we communicate has the power to shape the quality of our relationships. By listening with genuine curiosity and empathy, expressing ourselves authentically and validating others' experiences, we create spaces where people feel heard and valued. But communication isn't just about building connections – it's also a catalyst for change. When we communicate openly and honestly, we create environments where innovation can flourish, where barriers can be overcome and where progress can continue. Whether it's advocating for our own needs, motivating people towards collaboration and change or navigating conflict, effective communication is the driving force behind a sustainable workplace culture.

Personal Activity	*Taking A T-Break*
Objective	This tool, based on NVC techniques, encourages participants to consider the feelings and needs associated with experiences of conflict that are often missed when we focus too much on behaviours.
Duration	10–15 mins
Materials Needed	Piece of paper and a pen

Here's how it works:
Take a piece of paper and draw the following image:

Feelings | Needs

Figure 2.1 T-test.

(*Continued*)

Now think about a recent experience of conflict, either in your personal or professional life. Perhaps it was an argument, a disagreement or perhaps it was simply hearing someone say something that you didn't like hearing. Once you have brought an experience to mind, I want you to write down how you felt during that experience of conflict. Try to use emotion words.

Once you have written down some feeling words, I want you to identify what needs of yours were not being met during that experience of conflict? Was it your need to feel heard, respected, treated fairly? Write down as many as you can think of.

Once you have done that, draw another T with Feelings and Needs at the top. Now I would like you to imagine what you think the other person was feeling during that experience of conflict. Then what needs do you think they felt weren't being met or were perhaps trying to get met during that experience of conflict.

Take a look at your T's. Anything you notice. Are there similar feelings and needs being expressed, are there different ones? Often in experiences of conflict we are expressing the same needs and feeling the same things and yet will still manage to miscommunicate. The more we can be curious about the needs of the other as well as our own the more able to are to being the process of conflict transformation.

The T-Break is an example of Reflection-on-Action (more on this in the next chapter). It is something we can do after an experience to understand what happened better and do something different next time. The more you practice this skill, the more able you will be to move to the next phase, Reflection-in-Action, where you can do this in the moment, pause the conflict before it escalates and bring your attention back to the needs being expressed. The quicker you can identify and communicate needs, the quicker we can move to conflict transformation.

Team Activity	Mediation Workshop: Building Bridges to Conflict Transformation
Objective	To facilitate open dialogue, understanding and resolution of conflicts in the workplace through mediation techniques.
Duration	1.5 – 2 hours.
Materials Needed	Meeting room or designated space conducive to open discussion, Flip chart or whiteboard with markers, Mediation worksheets or handouts (optional).

Here's how it works:

Introduction (10 minutes): Introduce the purpose of the workshop. Set expectations of open communication, respect and confidentiality. Outline the agenda and any other ground rules.

Icebreaker Activity (15 minutes): Choose an icebreaker activity to help participants relax and begin build rapport with each other. Choose an activity that encourages sharing and collaboration, such as a team-building game or group discussion prompt.

(Continued)

Understanding Conflict (15 minutes): Facilitate a discussion on the nature of conflict, its causes and its impact on individuals and teams. Encourage participants to share their perspectives and experiences with conflict in the workplace. Use examples to illustrate common sources of conflict, such as miscommunication, differing expectations, or personality clashes.

Introduction to Mediation (10 minutes): Provide an overview of the mediation process, emphasising its collaborative and solution-oriented approach. Explain the role of the mediator as a neutral facilitator who guides the conversation and helps parties find common ground. Discuss the principles of active listening, empathy and respect in mediation.

Role-Play Exercise (30 minutes): Divide participants into pairs or small groups and assign roles (e.g., disputants and mediator). Provide a hypothetical conflict scenario relevant to the workplace (e.g., disagreement over project priorities or team communication issues). Encourage participants to role-play the scenario, with one person acting as the mediator and the others as the disputants. After each role-play session, facilitate a debriefing discussion to reflect on the experience, identify effective communication strategies and discuss potential solutions.

Group Discussion and Reflection (20 minutes): Facilitate a group discussion on the role-play exercises, focusing on what communication techniques were effective in resolving the conflict. Encourage participants to share their insights, challenges and lessons learned from the mediation process. Discuss how the skills and strategies learned in the workshop can be applied to potential conflict situations in the workplace.

Closing and Next Steps (10 minutes): Summarise the key takeaways from the workshop and reinforce the importance of constructive communication in conflict transformation. Thank participants for their participation and commitment to fostering a positive workplace culture.

Team Activity	Circle of Understanding: Exploring Perspectives in Conflict
Objective	To foster empathy, understanding and perspective-taking among team members to facilitate conflict transformation.
Duration	45–60 minutes
Materials Needed	Meeting room or designated space conducive to open discussion, chairs arranged in a circle, Flipchart or whiteboard with markers, sticky notes and pens.

Here's how it works:

Introduction (5 minutes): Introduce the purpose of the activity: to explore different perspectives in conflict situations and build empathy. Set expectations of open communication, respect and confidentiality.

Setting the Stage (10 minutes): Present a hypothetical conflict scenario relevant to the workplace, such as disagreement over project management approaches or team member roles. Briefly outline the different perspectives or interests involved in the conflict without assigning blame or judgment.

(Continued)

Sharing Perspectives (20 minutes): Invite participants to take turns sitting in the "hot seat" in the centre of the circle. Each participant in the hot seat shares their perspective on the conflict, expressing their feelings, needs and concerns. Encourage active listening and respectful engagement from the rest of the group as each person shares their viewpoint. Use sticky notes or a flip chart to jot down key points or themes emerging from the discussion.

Reflection and Group Discussion (15 minutes): Facilitate a group discussion based on the perspectives shared during the activity. What did participants learn from hearing different viewpoints, and how has it impacted their understanding of the conflict? Discuss common ground, areas of disagreement and potential solutions or compromises that could address the conflict.

Closing and Next Steps (5 minutes): Summarise the insights gained from the activity and emphasise the importance of empathy and perspective-taking in conflict resolution. Encourage participants to apply these skills in future conflicts and to seek opportunities for open dialogue and understanding. Thank participants for their participation and willingness to engage in constructive communication.

References

Deutsch, M. (2006). *The handbook of conflict resolution: Theory and practice.* John Wiley & Sons.

De Dreu, C. K. W., & Gelfand, M. J. (2008). Conflict in the workplace: Sources, functions, and dynamics across multiple levels of analysis. In C. K. W. De Dreu & M. J. Gelfand (Eds.), *The Psychology of Conflict and Conflict Management in Organizations* (pp. 3–54). Taylor & Francis Group/Lawrence Erlbaum Associates.

Fisher, R., Ury, W., & Patton, B. (2011). *Getting to yes: Negotiating agreement without giving in.* Penguin Books.

Miller, W. R., & Rollnick, S. (2012). *Motivational interviewing: Helping people change* (3rd ed.). The Guilford Press.

Rosenberg, M. B. (2015). *Nonviolent communication: A language of life* (3rd ed.). Puddle-Dancer Press.

3 Unconscious Bias: Developing Self-Reflection for an Inclusive Workplace

In the previous chapter, we explored how mastering the art of communication lays part of the groundwork for positive cultural change. We learned techniques for fostering deeper understanding between individuals and aligning teams around shared hopes and needs. However, even with strong communication skills, there are unavoidable psychological forces that can disrupt our best intentions. These come in the form of unconscious biases – ingrained stereotypes and prejudices that colour our perceptions and actions, often without us realising it. These blind spots are the by-product of how our minds simplistically sort and categorise the world based on what we can observe and our past experiences. Though usually unintentional, these implicit biases impact our decision making and communication and lead to an unhealthy culture of discrimination.

In cultivating an authentically inclusive environment, the role of self-awareness once again plays a pivotal role. We need to be able to take a step back and examine the lenses through which we view the world and question our automatic thought patterns. This requires a dedicated commitment to self-reflection, routinely checking our blind spots and catching biased perspectives before they negatively impact others.

If it were that simple, we would all be doing it. However, true self-awareness demands that we step outside our comfort zones, learn to question our underlying beliefs, shed light on our blind spots and assumptions and perhaps be introduced to some harsh truths about ourselves. Yet this inner work has profound reverberations for creating an organisational culture of equity, openness and mutual understanding.

Your Biased Brain: System 1 and System 2 Thinking

Your everyday thoughts, decisions and behaviours are being quietly influenced by unconscious biases. These biases hold significant power in shaping how we perceive others and navigate the complexities of our social environments. This unconscious refers to the ingrained beliefs and stereotypes that reside in the depths of our minds, influencing our perceptions and behaviours without us even realising it, and it is an often-overlooked phenomenon.

DOI: 10.4324/9781003407577-3

Despite our best intentions, these biases can slip into our decision-making processes, reducing equality, derailing inclusivity and hindering efforts at diversity. For an inclusive workplace culture to thrive, it is these three things we need more of, not less.

The concept of unconscious bias is rooted in social psychology and cognitive neuroscience. Early research in the mid-20th century shed light on the ways in which our minds process information and form judgements, revealing that our brains often rely on mental shortcuts and heuristics to navigate the complexities of the social world. Over time, researchers and scholars have expanded upon these findings, identifying the various forms of unconscious bias and exploring their implications for individual and collective behaviour (Gladwell, 2005).

Greater understanding of why this occurs has come from the field of Neuroscience, providing insights into the unconscious bias, highlighting the role of different cognitive processes in shaping our perceptions and decision making. The seminal work of Professor Daniel Kahneman *Thinking, Fast and Slow*, which emphasises how this occurs through the examination of what he calls the dual-process model of thinking, distinguishes between two main 'types' of thinking that people engage in: System 1, which is fast, intuitive and automatic, and System 2, which is slower, deliberate and analytical. Unconscious biases occur in System 1 thinking, where our brains rely on stereotypes to make rapid judgements without conscious reflection (Kahneman, 2011).

Okay, maybe, that's a bit too much neuroscience; let me give you a classic example. Answer this question as quickly as possible.

A bat and a ball cost £1.10. If the bat costs £1 more than the ball, then how much does the ball cost?

What was your answer, 10p? If so, that was incorrect; it's 5p. The majority of us will chose 10p because we use System 1 rather than System 2, instinct rather than analysis.

Imagine your brain as a complex computer, which it very much is, constantly processing your thoughts, ideas and perceptions of the world around you. This machine runs two key processing programmes, System 1 and System 2 thinking. System 1, the intuitive rapid-processing part of our brain, operates on autopilot. It's the force behind your split-second decisions and gut reactions. Imagine stepping out onto a busy street without looking, hearing a car and jumping back onto the pavement, you didn't stop to think about all the possible outcomes or reasons for the car sound, but you just reacted. That's System 1 at work. Conversely, System 2 is the deliberate, analytical processor. It's the mental software that comes online when we are faced with a complex problem to solve or find ourselves in an unfamiliar situation. It's a slow methodical programme that weighs up options and considers potential consequences before making decisions.

With the sheer amount of information your brain processes every second, unconscious bias is almost inevitable in the System 1 programme because it relies on our personal experiences and exposure to simplify and speed up decision making. This process or rapid decision making is helpful in some settings, allowing us to make decisions with minimal effort and less energy consumption; however, this increased efficacy has a cost, in that it is far more prone to errors in judgement and decision making.

Another way to look at it is the idea that your brain is generally not rational but instead it's rationalising. Answer these questions as quickly as you can, which do you prefer:

Apple or Android?
Netflix or The Cinema?
Sunshine or Snow?
Beaches or Mountains?
Sex or Chocolate?

The decisions you just made weren't rational. I imagine you arrived at your decision quickly. Now if I asked you to tell me why you made that choice, I'm sure you could come up with lots of reasons. However, in the first instance, you simply made a choice, engaging your System 1 process; it's only later that you bring your System 2 process online to rationalise the choice.

Let's imagine this in a social context. You've arrived at a networking event and you spot someone across the room who looks vaguely familiar. Your rack your brain trying to recall where you've seen them before. Suddenly, it hits you – they were at the same conference last year. With this realisation, your brain immediately starts rationalising why you should approach them:

> System 1 Thinking: "I remember them from the conference. They seemed friendly and approachable. I should go over and say hello".

Here, your brain quickly jumps to a conclusion based on intuition and past experience. It's a rapid, automatic response driven by System 1 thinking.

But wait, there's a twist. As you start walking towards the person, doubt creeps in. What if you're wrong? What if they don't remember you, or worse, they're not who you think they are? Suddenly, System 2 thinking kicks in:

> System 2 Thinking: "Wait a minute, let's think this through. Just because I vaguely recognise them doesn't mean they'll remember me. Maybe it's best to play it safe and wait for another opportunity to approach".

Now, your brain shifts into a slower, more deliberate mode of thinking. Instead of relying on instinct, you analyse and rationalise the situation carefully and consider the potential outcomes before taking action.

In this example, your brain's initial reaction was to rationalise approaching the person based on a hazy memory. However, upon closer inspection, System 2 thinking revealed the uncertainty underlying the decision. This illustrates how our brains can sometimes deceive us into thinking we're being rational when, in reality, we're simply rationalising our impulses and biases.

I imagine you can intuit how this type of biased thinking might influence decision making in the workplace. Imagine a hiring manager, tasked with selecting a new team member for a pool of candidates. Despite their awareness of the need to be unbiased and impartial, unconscious bias may seep into their decision making. Humans tend to gravitate towards similarity, and therefore, they may be more drawn to a candidate with a similar cultural, educational or professional background as them. Consider a supervisor conducting a performance review; their system 1 biases and preconception may influence them to rate certain employees more favourably than others. What about the fact that there are more CEOs called Andrew and Simon than there are female CEOs in the FTSE 100 (Bartram, 2023)? Are Simons and Andrews simply better CEOs? Perhaps some System 1 thinking is at work. In the workplace, unconscious bias can have significant implications for decision-making processes, hiring practices, performance evaluations and interactions among colleagues.

In order to start to challenge this bias, we need to accept that it is hardwired into our cognitive system and focus instead on developing our self-awareness and self-reflection in an effort to combat it.

Different Types of Bias

There are several flavours of unconscious bias that manifest in a number of different ways:

Implicit Associations: This refers to the unconscious connections our brains make between concepts such as race, gender or age and stereotypes or attitudes. These associations can influence our perceptions of biased judgements and decisions.

Workplace Example: Imagine a hiring manager reviewing CV's for a software engineering position. Despite efforts to remain impartial, the manager unconsciously associates certain names or educational backgrounds with specific demographics (remember our Andrews and Simons from earlier). As a result, they may unintentionally favour candidates with names or credentials that align with their implicit biases, leading to a lack of diversity in the final selection.

Confirmation Bias: This occurs when we selectively interpret information in ways that confirm our existing beliefs or stereotypes while disregarding evidence that contradicts them. In essence, we seek out evidence that supports our existing ideas, responding internally with "See I told you I was right!". This bias, in turn, reinforces our existing stereotypes and beliefs.

Workplace Example: A team leader assigns a project to a group of employees, including two individuals with different work styles. Over time, the leader notices that one employee consistently produces high-quality work, while the other tends to make mistakes. The leader subconsciously seeks out examples that confirm their initial impression of each employee, overlooking instances where the 'high-performing' employee makes errors and discounting successes from the other employee.

Affinity Bias: 'We like people like us'. Affinity bias involves favouring individuals whom we consider to be similar to us with regard to background, interests or values. We may be likely to treat those we perceive as similar more favourably and potentially disregard or devalue those we consider to be too 'different' from us.

Workplace Example: During a recruitment process, a hiring manager interviews a candidate who shares similar hobbies and interests. Despite other candidates demonstrating equal or superior qualifications, the hiring manager feels a sense of camaraderie with the like-minded candidate. As a result, they may unconsciously give preferential treatment to this individual, overlooking potential biases in the selection process.

Halo Effect: This occurs when our stereotypical impression of an individual influences our perception of their character, traits or capability. A classic example of this is when we perceive someone as physically attractive or likeable and therefore attribute them with other positive qualities such as intelligence, kindness or competence. The converse of this is something called Horns Effect, where we imbue people whom we find unattractive with a selection of negative traits or qualities such as laziness, incompetence or meanness.

Workplace Example: In a performance review meeting, a manager evaluates an employee who consistently arrives early, participates actively in team meetings and maintains a positive attitude. Due to these positive attributes, the manager perceives the employee as competent in all areas, including technical skills and problem-solving abilities, without thoroughly assessing each aspect of their performance.

Stereotyping: Relatively self-explanatory, this type of bias involves categorising individuals based on oversimplified and often inaccurate assumptions about their characteristics or abilities. Stereotypes are one of the most common causes of prejudice, discrimination and unequal treatment based on factors such as age, gender, race or ethnicity.

Workplace Example: In a corporate environment, a team leader assigns administrative tasks, such as note-taking and scheduling, to female employees, assuming they are more naturally suited to these roles. Conversely, technical assignments and leadership opportunities are predominantly given to male colleagues, based on the stereotype that men are more adept at analytical tasks and decision making.

Unbiasing Your Brain: Challenging Unconscious Bias

The antidote to unconscious bias is further developing the skill of self-awareness. The more you understand about yourself, the more you are able to notice your bias and challenge your assumptions. However, it can be hard to identify our unconscious bias directly, primarily because it is unconscious.

We usually only notice it after that fact, when we encounter a different opinion or when someone points it out to us.

Do me a favour, grab a piece of paper and a pen. On that piece of paper, I would like you to draw the following: A Table, a Clock and a House. Have a go now and keep reading once you've finished.

Ok. Take a look at your pictures. What did you draw? Did you draw an excel spreadsheet table or a football league table? I imagine the majority of you drew a rectangle with four legs, either two-dimensionally or three-dimensionally. If you did something else, bravo to you.

Next take a look at your clock. Did you draw the hands pointing at 3 o'clock? About 50% of people do. I'm not sure why; maybe because it was going home time at school, so it's drilled into your brain as the best time ever; maybe you just like right angles. Did you draw a digital clock? Probably not, you probably drew a circle with two hands. This is despite the fact that the clocks you see most often are the ones on your phone or computer, which are digital. Finally let me guess what you drew for your house. Are you ready? You drew a square, with a triangle on top, rectangular door, probably dead centre in the middle of the house. You probably drew square windows, and if you were feeling fancy, a little chimney on top with smoke coming out, a little path at the front and a tree to the left or right. Did you grow up in a detached cottage in the countryside; if not, why did you draw it?

I'll tell you why; schemas – the maps and models for stuff that you have in your head for what stuff is. You engaged your System 1 process and used a range of stereotypes and personal experiences to draw your standard model for what these things are. To put this into context, if you grew up in a house with a round table, you will probably draw a round table, and if you work in finance, you will probably draw and excel spreadsheet table.

You have these schemes for physical objects as well as concepts, what is a good person, a bad person, what is success and the like. They are not inherently bad on their own, just as System 1 thinking isn't inherently bad, but they can be unhelpful when we cling onto them rigidly with no room for variation. For example, I have a round table in my house. If I invited you in to come and sit at the table and you walked into the living room, would you say, "Wait there is no table in here; a table is a rectangle with four legs; what is this round monstrosity?"; no, you probably wouldn't. This is because you accept variation on your construct of table. However, you do this for some of your other schemas; you cling onto them rigidly with much room for variation. Mostly likely, however, you won't notice you have a rigid schema until you come across one that is different from yours and you have a reaction. So, you see, more often than not, it is when we encounter a different perspective that we have an opportunity to challenge our own.

That need not always be the case however; we can peer at some of our biases and schemes through increasing our self-awareness. One such method is to develop our reflective capacity through reflective practice.

Reflective Practice

The process of reflective practice is common in clinical and therapeutic settings, often overlooked in the workplace generally. It represents a structured approach to facilitate reflecting on experience. We had an example of this earlier in the use of the NVC T-Break activity. In brief, it is the process of exploring what happened, why it happened, how we felt about it and what we can learn from it to do things differently next time. It can be applied to specific situations or problems, and you can engage in it as an individual or facilitate a supported reflective discussion in a group.

Developing the skill of reflective practice and embedding it as a habit can support individuals and teams to:

- **Uncover New Information**: By exploring ideas and refining insights
- **Limiting Bias**: Through critical reflection and discussion of information
- **Building a Clearer Picture of the Situation/Process**: By exploring information, noticing contradictions and reaching a consensus.

The framework for reflective practice is modelled on something called the 'Experiential Learning Cycle', developed by Kolb (1984). It describes four stages in the process of reflection: Concrete Experience (Acting), Reflective Observation (Reflecting), Abstract Conceptualisation (Learning) and Active Experimentation (Planning). It offers a step-by-step process for learning from experience. Here is a simple breakdown:

- **Acting:** This is where we begin. It involves actively doing something or something having already happened. This may be something deliberate

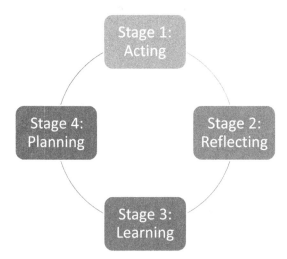

Figure 3.1 Flow diagram of four stages of reflective practice.

or unintentional, and it may have been 'successful' or not. Perhaps it's a new project at work, a challenging conversation or some incident that occurred.

- **Reflecting**: After the experience, you take a moment to pause and look back. You reflect on what happened, what you did and what you felt, re-examining the experience and making sense of it.
- **Learning:** Now, you can start making sense of your experience. You can try to understand the patterns, principles and concepts behind what happened. You can then use this awareness coupled with existing knowledge to plan for the future.
- **Planning**: Armed with your new insights, you're ready to return to the action. But this time, you hopefully approach the experience differently based on what you've learned. You try out new strategies, behaviours or approaches.

And then the cycle repeats itself. Each time you go through the cycle, you learn and grow. It's a continuous journey of learning and reflecting. The more you practice this skill, the better you get until you are able to do it in the moment rather than after the fact. This is the distinction between Reflection-on-Action, thinking after the event, and Reflection-in-Action, thinking while doing. The goal is to become good at Reflection-on-Action, catching our thoughts and reflecting on our experiences and behaviour as it is happening to allow us to make a more helpful and unbiased choice. The difference between asking at the end of the team meeting "I wonder what biases affected our decision making in this meeting today" and asking yourself that question as you are leading the meeting and challenging that bias before it leaves your mouth.

Unconscious Bias and Culture

Unconscious biases describe the ingrained stereotypes and prejudices that colour our perceptions and decisions outside of our conscious awareness. Though usually unintentional, these implicit biases breed discrimination and prevent the healthy cultural qualities of equity and inclusion from taking root.

Developing our self-awareness to a level where we can do this takes time but has a profound impact on the cultivating of a culture of authentic inclusivity. By embedding practices like reflective exercises, perspective-taking and the open discussion of biases, we can begin to develop the essential skill of reflection-on-action, preventing our unconscious assumptions from negatively influencing our culture. In doing this we can move more towards System 2 over System 1 thinking when it comes to key decision making. By understanding the processes by which we arrive at biased conclusions, we can do more to combat their presence in the workplace and begin to cultivate a culture of inclusion.

Challenging Bias: Practical Strategies for Increasing Reflecting Practice at Work

To effectively challenge unconscious bias, we need to develop our self-awareness and reflective capacity through consciously reflecting on our decision, behaviours and experiences to help gain insights and learning from them to do things differently in the future. In the workplace, reflective practice can be a powerful tool for professional development and critical thinking.

Team Activity	The Two-Speed Mind: Exploring System 1 and System 2 Thinking
Objective	To help participants recognise the difference between System 1 (intuitive, automatic) and System 2 (deliberate, analytical) thinking and understand how they influence decision making in the workplace.
Duration	1–1.5 hours
Materials Needed	Meeting room or designated space conducive to open discussion, Whiteboard or flip chart, Markers, Sticky notes.

Here's how it works:

Introduction (10 minutes): Introduce the purpose of the workshop. Set expectations of open communication, respect and confidentiality. Outline the agenda and any other ground rules.

Begin by introducing the concept of System 1 and System 2 thinking, providing a brief overview of each. Then divide the participants into small groups, ideally consisting of 4–5 members each. Distribute sticky notes and markers to each group.

Assign each group a workplace scenario where decision making is involved (e.g., choosing a candidate for a job, resolving a conflict, prioritizing tasks).

Idea Generation (20 minutes): Instruct each group to identify and list down the factors that might influence decision making in their assigned scenario. Encourage them to consider both intuitive (System 1) and analytical (System 2) factors.

Presentation (20 minutes): After generating ideas, have each group present their lists to the rest of the participants. Encourage discussion and reflection on how System 1 and System 2 thinking might play out in each scenario.

Group Discussion (20 minutes): Facilitate a group discussion on the potential biases or errors that can arise from relying too heavily on System 1 thinking, as well as the benefits of engaging System 2 thinking in decision-making processes.

Debrief (15 minutes): Invite teams to reflect on how they found the exercise, summarise key takeaways and explore how they can apply their understanding of System 1 and System 2 thinking in their everyday work tasks.

By engaging in this exercise, participants can gain a deeper understanding of how their thought processes influence decision making and learn strategies for mitigating bias and making more informed choices in the workplace.

Team Activity	*Exploring Personal Bias Through Storytelling*
Objective	To increase self-awareness and challenge unconscious biases by exploring personal narratives and experiences.
Duration	1.5–2 hours
Materials Needed	Meeting room or designated space conducive to open discussion.

Here's how it works:

Introduction (10 minutes): Introduce the purpose of the workshop. Set expectations of open communication, respect and confidentiality. Outline the agenda and any other ground rules. Gather participants in small groups of 3–5 individuals.

Story Sharing (25 minutes): Each participant takes turns sharing a personal story or experience related to diversity, inclusion or bias. The story can be positive or negative, but it should be authentic and meaningful to the individual. Encourage participants to focus on moments when they felt their perspectives or assumptions were challenged or when they became aware of their own biases, perhaps times where they felt included or excluded. Emphasise the importance of active listening and creating a safe, non-judgmental space for sharing.

Reflection and Discussion (35 minutes): After everyone has shared a story, allow time for group reflection and discussion. Prompt participants to reflect on the emotions, thoughts and reactions that arose during the storytelling process. Encourage them to consider how their own biases may have influenced their perceptions of the story.

Encourage open dialogue and exploration of different perspectives.

Identifying Patterns and Themes (20 minutes): After all participants have shared their stories, reconvene as a larger group. Facilitate a discussion to identify common patterns, themes and insights that emerged from the storytelling exercise. Encourage people to reflect on their own experiences and how they relate to the broader themes discussed.

Guide the group in exploring how unconscious biases may have influenced the stories shared and the participants' interpretations of those stories.

Action Planning (20 minutes): Conclude the activity by inviting participants to consider how they can apply their newfound insights and self-awareness to challenge unconscious bias in their daily lives. Encourage participants to set personal goals for increasing self-awareness and challenging bias, whether through ongoing reflection, seeking feedback from others or actively engaging in diversity and inclusion initiatives.

Team Activity	*Reflective Practice Session*
Objective	To develop the skill of reflective practice in teams.
Duration	1.5–2 hours

(*Continued*)

Materials Needed	Meeting room or designated space conducive to open discussion.

Here's how it works:

Introduction (5 minutes): Introduce the purpose of the workshop. Set expectations of open communication, respect and confidentiality. Outline the agenda and any other ground rules.

Select a recent work-related experience that team members can reflect on together. It could be a project completion, a challenging client interaction or a team meeting.

Guided Reflection (20 minutes): Facilitate the reflection process by asking open-ended questions that encourage reflection and self-awareness. For example, "What was your role in the experience, and how did you contribute?", "What were the key events or moments during the experience?", "How did you feel during the experience, and why?" and "What did you learn from the experience? How will you apply this learning in the future?".

Group Discussion (20 minutes): Encourage the group to share their thoughts and insights based on the questions with one another.

Identify Key Takeaways (15 minutes): Summarise the key insights and learnings that emerge from the discussion. Encourage participants to reflect on how they can apply these insights to future situations.

Action Planning (15 minutes): Finally, encourage participants to develop action plans based on their reflections. What specific actions will they take to leverage their learnings and improve their future performance?
By engaging in structured reflection sessions like this regularly, team members can develop their reflective practice skills and enhance their self-awareness as a way of challenging unconscious bias and developing towards the skills of Reflection-On-Action.

Team Activity	Critical Incident Reflection
Objective	To develop the skill of reflective practice through reflection on a specific incident. This can be used as a stand-alone session or as part of a Reflective practice Session.
Duration	1 hours.
Materials Needed	Meeting room or designated space conducive to open discussion.

Here's how it works:

Introduction (5 minutes): Introduce the purpose of the workshop. Set expectations of open communication, respect and confidentiality. Outline the agenda and any other ground rules.   .

Gather participants into one large group of a selection of small groups depending on the group size.

(Continued)

The What? (15 minutes): Offer a detailed description of the incident/ experience including the who, what, why, when and where.

So What? (15 minutes): Exploring the understanding people have of the incident and the sense they have made of it, including their perspective, emotions and actions.

Now What? (20 minutes): Draw connections from the experience/incident to identify further actions. What could you do differently next time? What are the lessons learned? How will we put new ideas or action into practice?

Looking Forward (5 minutes): Engage in action planning any Next Steps, how can we put these new ideas or perspectives into action.

References

Bartram, F (2023). "Average CEO in the UK", People Managing People. Available at: peoplemanagingpeople.com/news/average-ceo-uk/ (Accessed: 01 May 2024).

Gladwell, M. (2005). *Blink: The power of thinking without thinking.* Little, Brown and Company.

Kahneman, D. (2011). *Thinking, fast and slow.* Farrar, Straus and Giroux.

Kolb, D. A. (1984). *Experiential learning: Experience as the source of learning and development.* Prentice-Hall.

4 Motivation: Identifying Needs and Psychological Contracts

Greater self-awareness is usually a good thing, and just as it can help us identify our unconscious biases and assumptions, it can also help us to explore the factors that unpin not only our behaviour but also the behaviour of others. Motivation represents the 'why' that propels human behaviour, the combination of needs, values and beliefs with unconscious drivers that motivate people to act. These individual motivators are influenced by external situational factors such as the unspoken expectations and perceived obligations that form psychological contracts of a workplace culture. A culture that inspires long-term engagement and motivation is one that respects the diversity of individual motivations while seeking common threads to align those individual needs with the collective aims of the organisation.

Defining Motivation: The Driving Force Behind Behaviour

In a workplace context, motivation is the driving force that compels employees to take action, whether it's completing tasks, meeting deadlines or striving for professional growth. Motivation powers productivity and drives performance within an organisation, pushing sales teams to make sales and managers navigate obstacles, delegate tasks and keep morale high. It's what keeps teams focused on an 'end goal'. More often than not, you will only start to think about motivation when you notice it is lacking, when the team struggles to push through challenges, pursue new opportunities or miss targets.

From ancient philosophers to modern psychologists, we have been interested in the intrinsic driving forces behind human behaviour for centuries. Plato believed that humans were driven by a desire for harmony and balance in their lives, while Aristotle proposed that people sought to fulfil their innate potential and achieve eudaimonia – the pursuit of joy (Kraut, 2018). In the modern era, the model you are likely most familiar with is Maslow's hierarchy of needs, suggesting that humans are motivated by a hierarchy of needs ranging from basic physiological survival needs at the bottom, followed by safety needs, love and belonging, self-esteem and self-actualisation needs at the top. Just as we explored before, 'needs' form the basis for much of our

DOI: 10.4324/9781003407577-4

Figure 4.1 An example of Maslow's hierarchy of needs.

behaviour, as we attempt, through action, to get key needs met (Deci & Ryan, 2000).

Applying this to the workplace, we can use this model to explore and understand the needs and associated motivation of team members. For example, a new employee struggling with the 'cost of living crisis' may be primarily motivated by the need for a stable income and job security (physiological and safety needs), while a seasoned professional may be motivated by the desire for recognition and personal growth (self-esteem and self-actualisation needs), something we may likely see play out in their level of engagement at work. Understanding where employees fall on Maslow's hierarchy can help managers tailor their approach to meet their individual needs and foster a supportive work environment where they feel their needs are getting met. The higher up the hierarchy their needs are being met, the more connected, engaged and motivated they will feel.

Hopefully it seems obvious that for employees to even remain at the organisation, they need to feel their basic physiological and safety needs are being met – stable income, job security and the like. Although obvious, this is not always achievable; it is understandable that when an organisation goes through a period of 're-structuring' (business speak for people losing their jobs), engagement and motivation drop across the organisations, particular for those in the most insecure of roles. However, during periods of organisational stability, this is when it is important that we don't simply top at the first two but also that we consider the other needs in the hierarchy.

In a workplace context, 'love and belonging' refers to the psychological needs employees have of feeling a sense of connection to our colleagues and to the organisation. A sense of belonging is fundamental to our feelings or job satisfaction. Have you heard the phrase 'We don't leave organisations, we leave people'? It is often quality of the social connections at work that we take into account when we are deciding to leave or not. A loss of connection and belonging in organisations was one of the consequences of the pandemic and one we have yet to fully recover from. The more connected we feel, the more motivated we are and the more engaged we become.

Often, when organisations are trying to build motivation in employees, they skip right past love and belonging and head straight to esteem. Promotions, feedback, pay raises, awards – these are the go-to strategies for meeting employee esteem needs at work, and to a certain degree, they do work. Positive feedback and recognition go a long way to making employees feel valued and valuable. We all love a bit of dopamine, and being rewarded for 'good work' gives us just that. However, this is usually where organisations cease their efforts. Meet the basic needs, give suitable rewards to those who earn them and expect people to stay connected and motivated. However, this doesn't work in the long run. Think of all the jobs you have left: why you left, why you went elsewhere. Was it for more money (survival needs) or more advancement opportunities (esteem needs), or was it to do something that felt like it mattered, that matched better with your values and the things you were interested in doing (self-actualisation needs)?

The self-actualisation part of the hierarchy refers to finding a sense of meaning, purpose and fulfilment at work. Deep down, organisations appreciate the importance of this need; why else would your organisation have come up with a set of organisational values? I know you've got them. They probably represent lofty ideals such as justice, fairness, innovation and growth. When you can up with them, did they fix all your problems by the way? Were your staff suddenly filled with renewed motivation; did your retention skyrocket and your productivity double? Probably not. This is most likely because you came up with those values without actually finding out whether they aligned with the values of your workforce (we will explore this further in the next chapter). You can simply provide a set of meanings and expect people to agree with them. Our drive for self-actualisation is internal as well as external. If you can understand that, then you can help employees reach this vital part of the hierarchy, the thing that will really bind them to an organisation and motivate them towards shared goals. To understand this, we need to appreciate the difference between intrinsic and extrinsic motivation.

Motivating the Whole: Intrinsic and Extrinsic Motivation

We are most effectively motivated by a combination internal (intrinsic) and external (extrinsic) factors (Pink, 2009). Let me give you some examples:

Intrinsic Motivation: Consider a software developer who is passionate about coding and finds immense joy in solving complex problems. For this individual, the intrinsic rewards of creativity, intellectual challenge and mastery drive their motivation. They approach new projects with energy and drive and willingly invest extra time and effort to refine their skills because they derive genuine satisfaction from their work. Here they are motivation by internal needs for esteem and self-actualisation.

Extrinsic Motivation: Now, imagine a sales representative who is motivated by extrinsic factors such as monetary bonuses and recognition. This employee may be driven to achieve sales targets to earn commission or secure performance-based incentives offered.

However, on their own these factors may not be able to meet our needs in the long run. Intrinsic motivators are helpful because they encourage people to remain motivated even when external reward is lacking; however, they can be vulnerable to external factors changing, and if the external environment remains negative, our intrinsic motivators can begin to fade and shift and are often very hard to recover. Similarly, while extrinsic rewards like bonuses and praise can provide a short-term boost in motivation, they may not sustain long-term engagement if people lack the underlying intrinsic motivation.

Understanding the dynamic interplay between intrinsic and extrinsic motivation enables leaders and managers to design motivational strategies that meet both sets of needs simultaneously. By nurturing employees' intrinsic interests and values while providing appropriate external incentives, organisations can cultivate a motivated workforce that thrives on both personal fulfilment and professional success. For example, in a customer service role, an employee may find intrinsic fulfilment in helping others and making a positive impact on customers' lives. However, the company also offers extrinsic rewards such as quarterly performance bonuses and employee recognition programmes. Therefore, by aligning intrinsic values with external incentives, employees are more likely to feel valued, motivated and engaged in their roles. They may go above and beyond to deliver exceptional service, driven by both their internal sense of purpose and the tangible rewards offered by the organisation.

But do we want people to go 'above and beyond' and why would they do this? An intrinsic sense of purpose isn't usually included in their contract. This is because this is not the only 'contract' that influences employee motivation and behaviour.

Psychological Contracts

Perhaps you are familiar with this term, perhaps not. However, psychological contracts underpin our motivation and behaviours in all relationships, personal and professional.

They represent the unwritten, implicit agreements between employees and organisations, outlining mutual expectations, obligations and perceptions of the employment relationship. They exist whether you have thought about them or not. Unlike formal employment contracts, which usually specify extrinsic terms and conditions of employment such as mutual responsibilities, hours and pay, psychological contracts are subjective and based on the perceptions of 'promises' and 'commitments' made by both the employer and the employee. These contracts encompass not only the tangible extrinsic elements such as pay and benefits but also intrinsic and implicit factors such as workload, career advancement, recognition, support for wellbeing and work–life balance.

The concept of psychological contracts was first introduced in the 1960s. Initially focusing on the implicit expectations, contributions and reciprocations between employers and employees, it was eventually expanded by researchers including Rousseau (1989) and Guest (2004), to include broader societal and cultural influences on the employment relationship. Psychological contracts are dynamic and evolving constructs shaped by individual experience. Organisational practice and societal norms, the unspoken agreements, are often the silent contributors to workplace dynamics and, when left unexamined, can contribute to an unhealthy workplace culture. Let me share a couple of examples.

Consider the common scenario where an employee joins a company with the expectation of regular opportunities for skill development and career advancement based on informal discussions during the recruitment process. The employee perceives these 'promises' as part of the psychological contract, motivating them to commit their time and effort to the organisation. However, over time, the employee realises that these developmental opportunities are far more limited than they expected, and promotions seem to be based on unconscious bias and favouritism rather than merit. The employee experiences a breach of the psychological contract, leading to feelings of disappointment, disillusionment and reduced motivation.

Sometime the contracts themselves can directly contribute to an unhealthy culture. Picture a mid-sized recruitment company, where employees have an unspoken understanding that working long hours and sacrificing personal time are necessary for advancement. While this expectation is never explicitly communicated by management, it permeates across the entire organisational culture, leading to burnout, increased turnover and resentment among employees who have an intrinsic motivation towards prioritising work–life balance.

It is important that these implicit agreements are brought out into the open and healthy psychological contracts actively cultivated in order to maintain motivation in a healthy, productive workplace. If not, they can have a powerful influence on culture through:

Mismatched Expectations: When employees' implicit expectations do not align with the organisation's practices, they can feel disillusioned, frustrated or resentful. This, in turn, has a negative impact on morale and motivation.

Lack of Transparency: Unspoken psychological contracts throve in culture with poor communication. When organisations fail to communicate openly about expectations, employees may feel undervalued or even misled. This lack of transparency erodes trust, fosters cynicism and undermines motivation.

Perceived Inequality: Unchecked psychological contracts can lead to perceptions of unfair treatment, especially regarding the distribution of rewards, promotions or opportunities. These perceived inequities lead to resentment and polarisation and erode trust.

Lack of Psychological Safety: In an environment with consistently unspoken psychological contracts, psychological safety may be low, and employees may feel hesitant to challenge the status quo, voice dissenting opinions or propose innovative ideas for fear of reprisal or criticism.

Although often silent, psychological contracts exert a powerful influence on workplace dynamics and overall culture, shaping employee motivation, attitudes and behaviours. However, it is possible to shape these contracts for the mutual benefit of both the employee and the organisation, aligning expectations and motivations.

Shaping Psychological Contacts

In order to begin to shape psychological contracts, we need to start by noticing some of the powerful ones that are already there. These are often the behaviours and attitudes that are covered under the statement "The way we do things round here". You can start to explore them individually by asking yourself and your team to reflect on the expectations and perceptions of the employment relationship, such as development opportunities, wellbeing and organisational culture. What do you believe the organisation expects from you? What do you expect from the organisation?

You can do this one-to-one or gather feedback in the form or surveys, questionnaires or focus groups exploring perceived organisational promises and expectations. You can also look directly at your HR policies, leadership behaviours and broader organisation practices, to identify areas where there is a discrepancy between employee expectations and the organisational status quo. A classic example of this is the 'Wellbeing Policy' and states promises around work-life balance and wellbeing, coupled with the reality of leaders sending emails outside the working day and over the weekend.

Once you have identified some of the unhelpful silent contracts, you can begin to address these directly while putting onto practice strategies to foster healthy contracts in their place using the following framework:

1. **Promoting Trust and Transparency**: Encourage an environment of open communication where employees feel comfortable sharing their feelings, needs and goals. Clearly articulate your organisational and team values, goals and expectations to promote alignment between employees' individual goals and those of the organisation (the more in alignment, the more motivation). Finally demonstrate a commitment to fulfilling promises made to employees by following through on developmental opportunities, feedback and recognition. Creating a culture of transparency and trust allows positive psychological contracts to flourish.

2. **Mutual Accountability**: Psychological contracts are most effective when consciously co-created. Involve employees in the process of defining their roles, responsibilities, goals and performance expectations. Provide ongoing feedback to support employees to meet their goals and fulfil their commitments. Make sure that you align expectations with reality to prevent misunderstandings.

3. **Encourage Appreciation and Growth:** Make sure that you meet employees' esteem needs by recognising their contributions and achievements. This reinforces positive behaviours and strengthens their sense of value and belonging. Similarly invest in your employees' self-actualisation needs of growth and development by offering opportunities for them to fulfil their potential through training, mentorship. skill-building and career advancement. Supporting their personal growth in this way demonstrates your commitment to their success, meets those intrinsic hierarchical needs of esteem and self-actualisation and, in return, fosters loyalty and commitment.

4. **Cultivate a Positive Workplace Culture**: Encourage a culture of belonging where employees are supported to meet their intrinsic needs at work. Implement policies and practices that support employees' well-being, such as flexible work arrangements and wellness programmes. Challenge unconscious bias and embrace diversity and inclusion as core organisational values, celebrating employees' unique contributions.

5. **Lead by Example:** Managers and leaders are culture curators. Their behaviours set the tone for how employees are expected to behave, communicating many of the unspoken agreements that exist in an organisation. Lead by example by demonstrating healthy work practices, communicating with empathy, offering support and encouraging open dialogue and inclusivity.

6. **Monitor and Adapt**: Regularly check in on the health of the psychological contracts and address any emerging issues or concerns. Continue to evaluate policies and practise to continuously adapt changing needs and expectations of the organisation and the workforce. Finally, identify and address any perceived breaches in the psychological contract through open communication and collaborative problem-solving.

Consciously identifying and developing healthy and lasting psychological contracts between employee and the organisation can help in reducing turnover, improving wellbeing, fostering connection and maintaining motivation. Conversely, when left unspoken or disregarded, these unseen agreements can lead to disconnection, mistrust and disengagement.

Motivation and Culture

Motivation is what drives people to action. It is the combination of needs, values, beliefs and unconscious drivers that inspire people to do something and continue doing it. It is motivation that drives engagement, retention and productivity even during times of change.

At our core we are motivated by a set of universal needs, such as those outlined in Maslow's famous hierarchy: physiological requirements for security and stability, the yearning for belonging through quality social connections and the pursuit of esteem, mastery and meaningful impact. When these needs remain unfulfilled, apathy and disconnection thrive. But when they are met, employees become energised and dedicated to the task at hand.

Of course, individual motivations don't emerge in a vacuum. They interweave with the implicit psychological contracts silently permeating the workplace, the unspoken reciprocal obligations and expectations regarding effort, development opportunities, work–life balance and more. These usually unconscious agreements exert a powerful influence on the dynamics and health of an organisation's culture. Misaligned expectations erode trust and a sense of inequality. Rigid, unhelpful, cultural norms instil fear of going against the status quo. But by acknowledging psychological agreements and aligning them with employee and employer needs alike, we lay the groundwork for healthy, sustained motivation.

In essence, we need to begin to rehumanise the workplace by intentionally designing it around what ignites intrinsic inspiration and the extrinsic incentives that help maintain it. Though the alignment of inner and outer drivers, employees are able to feel that their needs are being met and therefore willingly invest their skills and passion in service of a vision they feel a part of. What is the pin that holds it all in place? The construct that solidifies the alignment of motivation and vision? Values (something we will delve into in the next chapter).

Motivational Contracts: Practical Strategies for Identifying Personal Motivators and Influencing Psychological Contracts

To cultivate a culture of safety, belonging and alignment, where people motivated to contribute their best work, we need to explore the intrinsic and extrinsic motivations of employees. Once you understand what motivates your employees and the psychological contracts that maintain those motivations, you can actively engage in the construction of healthy, sustainable psychological contracts to maintain that motivation in the long term.

Team Activity	Motivational Interviews
Objective	The aim of this activity is to support leaders to discover what truly motivates their staff through one-on-one motivational interviews or meetings and creating a space for employees to express their thoughts, feelings and aspirations regarding their work and career.
Duration	1 hour
Materials Needed	Meeting room or designated space conducive to 1:1 discussion.

(Continued)

Here's how it works:

Schedule individual meetings: Arrange private meetings with each team member to discuss their motivations and goals. Ensure that the meetings are conducted in a confidential and non-judgmental manner to encourage honest communication.

Use open-ended questions: During the meetings, use motivational interview strategies such as open-ended questions to encourage employees to share their thoughts and feelings. You might ask:

- *What aspects of your work do you find most fulfilling?*
- *What are your long-term career goals?*
- *Can you describe a time when you felt particularly motivated or engaged at work?*
- *How do you prefer to receive feedback and recognition?*

Practice Active Listening: Practice active listening during the meetings, paying attention to verbal and non-verbal cues. Show empathy and understanding towards employees' experiences and perspectives.

Explore underlying needs: Seek to uncover employees' underlying needs and values. For example, if an employee expresses a desire for career advancement, inquire about the specific aspects of advancement that are important to them, are they seeking learning opportunities, recognition or leadership roles?

Take notes and follow up: Make sure to take note of key points during the meeting to identify employees' motivations and aspirations. Follow up with employees regularly to check in on their progress and ensure that their needs are being met.

By engaging in motivational interviews to identify individual needs and motivations, leaders can gain insights into what drives and inspires their team members. This information can then be used to tailor leadership approaches, assignments and opportunities to better align with employees' motivations, leading to increased engagement, satisfaction and productivity in the workplace.

Team Activity	*Motivation Reflections*
Objective	The aim of this activity is to help employees identify and reflect on their intrinsic and extrinsic motivators, fostering self-awareness and engagement in the workplace.
Duration	1–1.5 hours
Materials Needed	Meeting room or designated space conducive to open discussion, Flip chart paper or whiteboard, paper and Pens.

Here's how it works:

Introduction (10 minutes): Introduce the purpose of the workshop. Set expectations of open communication, respect and confidentiality. Outline the agenda and any other ground rules.

Provide employees with self-reflection prompts, containing prompts to explore their motivations. Include questions such as:

- *What aspects of your job do you find most fulfilling and enjoyable?*
- *What tasks or projects energise you and make you feel accomplished?*

(Continued)

- *Do you have any long-term career goals or personal aspirations related to your work?*
- *Are there any external rewards or incentives that influence your performance or engagement at work?*

Group Discussion (20 minutes): Split the team into small groups and facilitate a group discussion or where employees can share their reflections on the questions in a supportive and non-judgemental environment.

Identify Themes (15 minutes): Encourage groups to identify common themes or patterns in their responses. Are there recurring intrinsic motivators such as autonomy, development or purpose? Do they notice any significant extrinsic factors such as financial incentives, recognition or career advancement? What about intrinsic factors such as personal growth, purpose or fulfilment?

Goal Setting (25 minutes): Based on their reflections, encourage employees to set personal goals aligned with their intrinsic values and aspirations. Encourage them to identify concrete actions they can take to leverage their intrinsic motivators and mitigate any barriers to motivation. Ask them to identify specific steps they can take to leverage their motivations and strengths in their current roles, as well as opportunities for growth and development that align with their values and aspirations. Invite them to share these personal goals with the rest of the group.

Debrief (15 minutes): Encourage group to reflect on their experience of the exercise, what they noticed and what they have learned. Encourage the group members to arrange meeting with their line manager to share their goals and motivating factors. Encourage them to review their progress towards their gal with their line manager, reflecting on, adapting and refining motivations as they go. This exercise encourages self-awareness and mutual accountability for motivation among the team supporting them to identify and leverage intrinsic and extrinsic motivators.

Team Activity	Identifying Implicit Contracts
Objective	The aim of this activity is to uncover some of the implicit expectations and commitments that exist between employees and the organisation.
Duration	1–1.5 hours
Materials Needed	Meeting room or designated space conducive to open discussion, Flip chart paper or whiteboard, Markers, Sticky notes and Pens.

Here's how it works:

Introduction (10 minutes): Introduce the purpose of the workshop. Set expectations of open communication, respect and confidentiality. Outline the agenda and any other ground rules.

Start by explaining the concept of psychological contracts and their significance in the workplace. Emphasise that these contracts represent the unwritten, implicit agreements between employees and the organisation regarding expectations, contributions and rewards.

Then divide participants into groups of 3–5.

(*Continued*)

Idea Generation (15 minute): In their groups ask them to identify and list down the implicit expectations they believe employees have from the organisation and vice versa. Encourage them to consider various aspects such as career development, wellbeing, work-life balance, recognition and organisational support. You might ask question like;

- *What do I expect from the organisation in terms of career development, recognition, support and work-life balance?*
- *How do I perceive the organisation's expectations of me regarding performance, commitment and contribution?*
- *Are there any unspoken assumptions or beliefs influencing my behaviour and attitudes towards the organisation?*

Sharing & Discussion (20 minutes): After all groups have shared their expectations, facilitate a discussion to identify common patterns and themes. Encourage participants to reflect on the similarities and differences in expectations, noting any areas of alignment or potential discrepancies.

Reflection (15 minutes): Distribute sticky notes and pens to each participant. Ask them to individually reflect on the following questions:

- *What are the most important expectations you have from the organisation?*
- *How do you believe the organisation perceives your role and contributions?*
- *Identify the areas where your expectations align or conflict with those of the organisation?*

Group Discussion (15 minutes): Invite participants to share their reflections with the group, sticking their notes on a designated area of the wall. Facilitate a discussion around the shared reflections, exploring any common themes or insights that emerge.

Debrief (10 minutes): Conclude the activity by summarising the key takeaways and insights gained from exploring the psychological contracts between employees and the organisation. Explore how they can apply this knowledge to developing healthy and sustainable workplace contracts.

References

Deci, E. L., & Ryan, R. M. (2000). The "what" and "why" of goal pursuits: Human needs and the self-determination of behaviour. *Psychological Inquiry, 11*(4), 227–268.

Guest, D. E. (2004). The psychology of the employment relationship: An analysis based on the psychological contract. *Applied Psychology, 53*(4), 541–555.

Kraut, R. (2018). Aristotle's Ethics. In E. N. Zalta (Ed.), *The Stanford Encyclopedia of Philosophy* (Summer 2018 ed.). Stanford University.

Pink, D. H. (2009). *Drive: The surprising truth about what motivates us.* Riverhead Books.

Rousseau, D. M. (1989). Psychological and implied contracts in organisations. *Employee Responsibilities and Rights Journal, 2*(2), 121–139.

5 Values: Leveraging the Power of Meaning and Purpose

At the core of the majority of organisational leadership theories and practices lies the concept of values. These describe the deeply held beliefs that guide our decisions and actions. They form the cornerstone of our personal identity and naturally have a significant impact on our motivation and the workplace ecosystem.

Our values act like a compass guiding us towards our self-defined 'purpose' in life. They are at the heart of what we do, our motivation, behaviour, satisfaction, resilience and wellbeing. The study of values originates once again with Aristotle, the most prolific of thinkers, in his ponderings on the nature of virtue and the pursuit of eudaimonia (meaning, purpose, joy). Over time, a number of psychological theories and frameworks have developed to help define our understanding of the nature and development of our core values (Schwartz, 1992).

Let's think about yours for a second. Your values are a fundamental part of who you are or who you want to be. They define the things that you believe are the most important in life. They determine your priorities and are the yardstick by which you measure whether your life is turning out as expected. They are usually pretty stable, although they may change as you move through life. For example, at the start of your career, you might value status and success, and then later if you decide to start a family, your values may shift to things like work-life balance. Have you got an idea of what some of your values might be yet? Try this exercise.

Values Self-Assessment

- I want you to think of a time when you were happiest. It can be an example from both your personal life and your work life if you want. When you have brought a specific moment to mind, ask yourself the following questions and note down your answers:

  - *What were you doing?*
  - *Were you with other people? If so, then who?*
  - *What factors contributed to your happiness in that moment?*

DOI: 10.4324/9781003407577-5

- Now think about a time you were most proud. Again, you can use examples from both work and personal lives. When you think you have something, ask yourself the following:

  - *What were you doing?*
  - *Did other people share in your pride in that moment? If so, then who?*
  - *What factors contributed to your feeling of pride?*

- Finally, think of a time where you felt most fulfilled. Again, use examples from both work and personal lives. When you have identified a moment, ask yourself the following:

  - *What were you doing?*
  - *What need or needs of yours were being met in that moment you felt fulfilled?*
  - *What factors contributed to your feeling of fulfilment in that moment?*

- Take a look at your answers to each of the above. Try to draw out any themes or similarities across the three question sets. Start to distil your answers to come up with core values, things that mattered to you in that moment. Some example of core values are as follows: Integrity, Compassion, Community, Authenticity, Achievement, Challenge, Fun, Learning, Peace, Recognition or Service. Try to combine similar values until you end up with a small set of core values that feels comfortable.

What do you think? Do these represent your core values? Are they the ones you would have said if I asked you directly? Often, when I ask this question to groups "What are your core values?", I get very noble examples. 'Integrity', 'Justice', 'Fairness', 'Respect', 'Honesty' and the like. Once I receive these, I usually follow it up by asking people to grab a piece of paper and break down their day into 1-hour increments from the moment they wake up to the moment they go to sleep. You can try it yourself now if you like. Once they have the different time slots, I ask them to write down what they usually do in those different time slots from the start to the end of the day. Then I ask them to take a look and what they haven't written down and ask themselves whether those actions are all. 'Integrity', 'Justice', 'Fairness', 'Respect' and 'Honesty'? For many of us, they are things like eat breakfast, scroll Instagram, send emails, eat my lunch at my desk, have another coffee, TV and then bed. Maybe they simply value those things, and that's ok; you can also choose different values like Instagram and coffee if you want. Otherwise perhaps you may need to notice that what you are doing each day doesn't actually align with your values and maybe that's why you don't feel very happy.

Now in reality it's usually a little bit of both. Sometimes all the tedious actions that don't necessarily bring us joy are still necessary in pursuit of our actual values such as freedom or success. The important thing to notice is that when we live a life that is in alignment with our values, we feel good (note your responses to the values of self-assessment above). Conversely, when we live a life that is not in alignment with our values, we tend to feel disconnected, disengaged, unmotivated and low. If we switch the exercise questions above and I ask you to think of a time you felt most unhappy, most disappointed and most dissatisfied and explore what was happening and contributing to that feeling in that moment, I imagine you will notice that those moments did not align with your values. Your values provide you with a sense of meaning and purpose; they are the route to the top of Maslow's hierarchy, self-actualisation.

In the workplace, core values have a significant impact on:

Workplace Culture: Serving as the cornerstone of your culture, guiding behaviour, decision making, social dynamics and a sense of belonging and shared purpose.

Leadership Approaches: Defining how leaders lead, in terms of integrity, authenticity and ethical decision making. Those who embody shared values tend to encourage a culture of trust, accountability and safety.

Decision Making: Informing strategic decision making at work with regard to guiding choices that either align with the organisational purpose and mission or don't. When faced with difficult decisions or dilemmas, an organisation can use their core values as a shared set of guiding principles.

Moreover, just as you can support them to meet their core needs, you can support employees to get some of their core values met at work. You can actively cultivate a workplace wellbeing culture where people feel motivated and engage without the need for constant 'Stress Management', 'Motivation' and 'Wellbeing' training programmes.

The Power of Meaning and Purpose

Fostering a sustainable workplace culture is essential to your organisational success. To be sustainable it needs to be self-perpetuating, where the culture is structured in such a way that it encourages greater organisational wellbeing. One way in which we can achieve that is leveraging the power of 'Meaning' and 'Purpose' in the workplace.

Just as meaning and purpose can offer us a sense of direction and satisfaction in life, they can offer us the same things at work. Employees who find a sense of meaning at work usually experience greater job satisfaction and therefore be more engaged and productive. Sharing a sense of purpose at work can also foster greater team cohesion, connectedness and loyalty to an organisation. These all sound like pretty helpful things, no? But how do we leverage this at work? Let me offer some simple examples.

Imagine a non-profit focused on environmental conservation, which inspires its employees by directly connecting the work they do to a larger mission of protecting and preserving the planet for future generations. As a result, employees feel a deep sense of purpose and a belief that their work matters, leading to increased engagement and commitment.

Alternatively, what about a housing company which puts aside a portion of its profits to support local housing initialise and encourage employees to also volunteer their time and skills to these initiatives? By aligning the employees' skills and work to direct social impact in their local communities, they feel a sense of meaning and purpose to what they do, increasing their level of satisfaction and commitment.

However, not every organisation is in the position to support charitable initiatives like this, and giving back may not be a value that everyone in the organisation shares. Therefore, a more broad application approach to harness the potential of meaning and purpose can be found in values-based leadership.

Values-Based Leadership and the Myth of Organisational Values

Being aware of and leading from a position inspired and directed by values represents a shift away from traditional leadership perspectives towards the concept of values-based leadership. Here, rather than being solely focused on outcomes and strategies, leaders place an emphasis on aligning leadership actions with their authentic personal values as well as identifying and aligning actions with employee values. The values-based leadership perspective recognises that when we lead from a place of authenticity and integrity, and support people to be in alignment with their values, we are far better able to inspire and motivate employees (Brown, 2018).

Let's start with what happens when organisations are not in alignment, as this is likely what you are more familiar with. Firstly, let's acknowledge the attempts to harness meaning and purpose through the use of organisational values. These have been increasingly popular every year. In fact, every organisation I talk to now seems to have a set of organisational values; unfortunately, for most organisations they don't know why they need them, just that they need. I have seen these lists of lofty ideals adorn email signatures, conference room walls and employee handbooks, but something has usually been missed; otherwise, so many wouldn't feel like empty slogans more than guiding principles.

In a scene common to many of us, picture yourself walking into the office lobby after the weekend to be greeted by a display of colour posters that have appeared seemingly overnight, adorned with words like 'Integrity', 'Innovation' and 'Excellence'. Your new company values have arrived. The handbook has been updated, an email has been sent out to everyone and you've been invited to a workshop to talk about how you can live by these values at work. Yet amid all the motivational quotes and corporate sounding buzzwords, you can't help but wonder whether these values actually mean anything, if anything has actually changed other than the decoration on the walls. Sure, the words sound impressive, but that's usually what buzzwords do; they sound visionary and forward-thinking, but in reality, they often lead to confusion and a sense of misalignment when we struggle to relate to these 'values' we've been told to share. Perhaps this is not the case in your organisation; perhaps you understand how company values can be powerful, and as a result, you developed them after holding workshops and surveys across your entire workforce, distilling them down to a total of three values. If no, then keep reading.

When Our Values Don't Align

Companies often spend so much time, and money, coming up with values to bind a workforce together and reap the benefits that they overlook the impact of misaligned values in an organisation. Sometimes, their strategies to align values backfire directly to misalignment. Misaligned values occur when there is a disconnect between an individual's or organisation's stated values and their enacted values, demonstrated through the actual behaviours, beliefs and decision making in the organisation. So how can you spot a misalignment? Look out for these:

- **Lack of Engagement and Connection**: Employees may feel disengaged or withdrawn when their values do not match the values stated or enacted by the organisation, impacting their emotional engagement with work. You might see higher turnover, absenteeism and low wellbeing.
- **Loss of Trust**: Values misalignment can erode trust across the organisation, which, in turn, can affect collaboration and teamwork. You might see more siloing, increased conflict or a drop in psychological safety.
- **Loss of Motivation:** When employees feel that their values are not aligned with an organisation's goals, they are less motivated to perform at their best. You might see decreased job satisfaction, lower commitment and reduced effort.
- **Resistance to Change:** When company values are misaligned, and employees do not feel connected to a shared sense of purpose, they may begin to resist change or innovation. Why follow you if they don't know where you're going? You might see reluctance to let go of outdated practices, rejection of new initiatives or technology and less resilience and adaptability.
- **Negative Overall Culture:** Values misalignment tends to lead to a negative overall culture across the organisation, one that is usually characterised by cynicism, distrust and conflict. You might notice all of the above plus a poor organisational reputation (check sites like Glassdoor.co.uk to see what yours is), difficulty with recruitment and retention and stifled growth.

If you can start to notice the signs of misaligned values and call out the elephant in the room of your meaningless 'organisational values', then you can actually take some proactive steps to begin to make the most of meaning a purpose and re-align your people's and your company's values.

Re-Aligning Values and Encouraging Meaning and Purpose

You can begin a process of re-alignment through your leadership style as well as across the organisation as a whole. Enacting values-based leadership means bringing awareness to the alignment between your actions and decisions with your core values and those of the organisation. Talking the talk and walking the walk of values so to speak. Leaders who use their values as a compass to guide their actions tend to demonstrate the following:

- **Integrity**: A leader demonstrates integrity by keeping promises, taking responsibility, communicating with transparency, being honest and accountable and adhering to a strong set of ethical standards.
- **Empathy**: A leader shows empathy through active listening, acknowledging feelings, considering the impact of their actions and offering support and understanding during challenging times.
- **Collaboration**: A leader encourages a culture of collaboration through open communication, cross-functional collaboration activities, breaking down silos and encouraging knowledge sharing to increase creativity and innovation.
- **Respect**: A leader demonstrates respect by actively seeking employee input, valuing individual contributions, giving constructive feedback, treating everyone with dignity and fairness and fostering an inclusive work environment.
- **Accountability**: A leader holds themselves and their team accountable for providing clear expectations, achieving goals, addressing performance issues promptly, learning from failures and demonstrating a commitment to growth.
- **Authenticity**: An authentic leader is aware of their values and able to communicate them, is sincere, is genuine in their communication and inspires trust and connection among the team.

Many of these skills we have covered in the earlier chapters, exploring their relationship to the cultivation of a sustainable workplace culture. They are no less important here in being able to align your values with your actions as a leader and demonstrating behaviours that employees can align with their own values. As an organisation, in order to encourage more alignment than misalignment, you can do the following.

Clarify Your Organisational Values and Purpose

It is important that you clearly communicate the organisation's values and purpose (often called a 'mission statement') to your employees, helping them understand how their work contributes to larger goals of the organisation. Step one: come up with some values that actually make sense and genuinely align with your workforce and what you are collectively trying to achieve.

I have given an example of an exercise to do this at the end of this chapter; however, in brief, you can ask them things like, What they think is important to the organisation, what is important to them about working for the organisation and why they think their work matters? Step two: once you have collected feedback from across the organisation, start to distil it down until you arrive and what feel like some core values. Step three: share them again with the organisation to get their feedback on the final selection and whether they feel right. A quick word of warning here about number of values. Too often do I see organisations with lists of values that are five, seven, ten values long. Stop it. The minute I see more than four, it's clear to me that no expert consultation, or even internet consultation, was involved in the decision to choose those values. As a basic rule of thumb, if you have more than four values, start again. This relates primarily to our capacity to generally only be able to hold four concepts in our mind at once (Cowan, 2010). If you are giving your employees more than four values to pay attention to, you have already lost them. My general rule for organisational values: four is ok, three is ideal!

Once you have clarified your organisation's real values, come up with a mission statement that draws those values together into a raison d'être, a reason for being. Keep it short and snappy and speak to the value you bring and the change you make in the world. I have described an exercise at the end of this chapter to come up with both a team and an organisational purpose statement. Use something like this to make sure your purpose statement is accessible and simple and genuinely connects to your employees' values. To give you an example from my work, I once worked with a housing company to help them devise a purpose statement, and after a few workshop sessions, we arrived at "We make dreams come true. We help people by their first home and live the life they choose". This was a much more powerful motivator for people than "We sell mortgages" or "We build houses". If you'd like a real-world example consider outdoor clothing Patagonia's mission statement – "We're in business to save our home planet" (Chouinard, 2016). Now that is a mission statement people can rally behind. More on Patagonia's approach to culture in Chapter 8.

Foster a Culture of Opportunity, Appreciation and Teamwork

Be conscious of those upper levels of Maslow's hierarchy of needs, and provide opportunities for personal growth and development through things like mentorship, training and clear pathways for career advancement. Importantly, make sure access to these opportunities is as unbiased and transparent as possible. Most of us harbour a value of fairness (often called a just-world belief) when we feel it has been broken or unmet; we quickly lose trust in the individual or organisation who broke it, and it is often very hard to regain that trust.

Actively recognise and celebrate employees' contributions and achievements. Help them see the value in their work by finding ways to reinforce the direct relationship between their work and its impact on the organisation. How does the work they do contribute to the greater purpose of the organisation? In the example of the housing company I gave above, even the company finance manager believed that they too 'make dreams come true', because they could see directly how their efforts helped other people succeed in theirs. In the Patagonia example, employee actions are linked directly to global conservation efforts (Chouinard, 2016).

If you can demonstrate how their work, no matter what it is, has a greater meaning, they will feel more connected and aligned to organisational efforts and be motivated not only by the greater purpose but also by working in tandem with their colleagues. This can work both ways. If I'm not a fan of teamwork but I value helping others, I am more likely to engage with teamwork activities if I understand how it serves this greater purpose.

Support Work–Life Balance

I sincerely doubt that anyone on your staff team values 'doing a 9–5'. They might value what that gives them, such as security, providing for their family, challenge, growth or freedom. Therefore, find ways to demonstrate that you value their life outside of work as much as you value it in work. Because everything that happens outside work will affect how they perform in work.

If you haven't already, make sure employee wellbeing or happiness as part of your core company values. Because you do value it whether you believe you do or not. Maybe you think you value productivity and innovation, but both are consequences of a happy, connected staff team, so deep down you value this more. But don't just say it; live it. Communicate that you prioritise employee wellbeing by offering classic incentives such as flexible working for everyone. Design and deliver thoughtful wellbeing programmes that align with what your staff team considers to be stuff that supports their wellbeing; don't offer them a day where you fill one of the offices with puppies when they are all cat people. Provide effective and accessible resources for managing stress and burnout. The only constant in life is change, and with change often comes stress. Think about what you currently offer staff to support them to manage stress, stress that will be caused in part by the demands placed on them at work. Is your current strategy free coffee and biscuits? Ah yes, caffeine and sugar; great for stress management. Again, ask them what they want, what would meet their wellbeing needs. Did they ask for a 'Chill-out Room' or did you get one because Google has one and you thought it was cool? Give them what they need, not what you think they need; they are the experts in them. The more you align your initiatives with their needs, the more aligned and connected they will feel.

Values and Culture

To cultivate a culture of alignment, engagement, wellbeing and sustained effort, you need to start harnessing the power of values, meaning and purpose. Doing so will help meet the intrinsic motivations of your staff and provide a sense of connection to the organisation and its overall goals. This doesn't happen when values feel incongruent or merely symbolic. Values can provide a compass for guiding behaviour, decision making, leadership integrity and collaborative momentum, but they must ring true and, where possible, emerge as an organically organisational system itself. If you can demonstrate how each person's contribution genuinely contributes to the overall aims of the organisation, you can align opportunities for growth and impact with their individual hopes and needs. This values-based approach to culture nurtures mutual understanding, trust and goodwill.

Understanding the importance of values and talking about them aren't enough however; you need to embed them into every facet of the business and how it operates. This requires leaders developing the self-awareness to walk the talk and modelling vulnerability, accountability and ethical decision making. It requires designing processes, incentives and interpersonal dynamics to reinforce your professed cultural values day-to-day. Yet when we get it right, we support the growth of a resilient culture of shared commitment and sustained effort, one where improved wellbeing and performance are the consequences of work imbued with purpose. We all want to lead meaningful, purposeful and fulfilling lives, and there is no reason that we cannot find this meaning at work.

This individual and organisation resilience that is bolstered by the alignment of values is necessary for navigating the challenges that will inevitably arise in the journey of any organisation.

Values-Based Culture: Practical Strategies for Identifying Personal and Organisational Values and Defining a Sense of Purpose

In order to leverage the power of meaning and purpose in a truly values-based organisation, it is important that you identify what your employees' and company's values actually are and discern how to meaningfully embed and enact these to curate a shared sense of purpose.

Team Activity	Uncovering Core Values
Objective	The objective of this exercise is to help individuals identify their core values by reflecting on their beliefs, priorities and aspirations. By gaining clarity on their values, participants can align their actions and decisions with what truly matters to them, leading to greater satisfaction and fulfilment.
Duration	1–1.5 hours
Materials Needed	Meeting room or designated space conducive to open discussion, Flipchart paper or whiteboard, paper and Pens.

Here's how it works:

Introduction (10 minutes): Introduce the purpose of the workshop. Set expectations of open communication, respect and confidentiality. Outline the agenda and any other ground rules.

(Continued)

Begin by explaining the purpose of the exercise: to explore and uncover personal core values. Emphasise the importance of values in guiding decision making, setting goals and living authentically. Encourage participants to approach the exercise with an open mind and to be willing to share their thoughts and feelings. Split the team into smaller groups of 4–6 people.

Reflection (15 minutes): Ask participants to take a few moments to reflect silently on the following questions, writing down their thoughts and feelings on paper, without overthinking or censoring themselves.

- *What principles or beliefs do I hold most dear?*
- *What aspects of life bring me the greatest sense of fulfilment and joy?*
- *When do I feel most aligned with my true self?*

Group Sharing (15 minutes): Invite participants to identify common themes and insights that emerged from the reflections and share them with the wider group. Encourage active listening and validation of each participant's experience.

Values Identification (20 minutes): Using the insights gained from the reflection exercise, guide participants through a process of identifying their core values from the themes they noticed. You can assist by providing a list of common values (e.g., honesty, compassion, resilience) and asking participants to choose the ones that resonate most deeply with them.

Encourage participants to rank their chosen values in order of priority, based on how strongly they feel about each one.

Values Clarification (15 minutes): Facilitate a discussion around the chosen values, exploring why each participant resonates with them. Encourage participants to reflect on how their values align with their life choices, goals and relationships. Prompt participants to consider any conflicts or discrepancies between their stated values and their actions or experiences.

Action Planning (10 minutes): Encourage participants to identify specific actions they can take to be more in alignment with their core values in their daily lives. Emphasise the importance of setting concrete goals and holding oneself accountable for living in alignment with one's values. Provide a space for participants to share their action plans with their small groups or with the group as a whole, if they wish to do so.

Reflection and Closing (5 minutes): Invite participants to reflect on what they've learned from the exercise and how they plan to integrate their core values into their lives moving forward. Offer words of encouragement and support as participants embark on their journey of living authentically and aligned with their values.

Team Activity	Identifying Organisational Values
Objective	The objective of this exercise is to help organisations identify their core values in a meaningful and effective way. By engaging stakeholders in a collaborative process, the exercise aims to uncover shared beliefs, priorities and aspirations that define the organisation's identity.
Duration	1.5–2 hours

(Continued)

Materials Needed	Meeting room or designated space conducive to open discussion, Flip chart paper or whiteboard, sticky notes, and markers.

Here's how it works:

Introduction (10 minutes): Introduce the purpose of the workshop. Set expectations of open communication, respect and confidentiality. Outline the agenda and any other ground rules.

Begin by explaining the purpose of the workshop: to identify and define the core values of the organisation. Emphasise the importance of values in shaping organisational culture, guiding behaviour and fostering alignment among stakeholders. Set clear expectations for the workshop, and encourage participants to actively participate and contribute their perspectives.

Idea Generation Session (20 minutes): Divide participants into groups of 4–6, representing different departments and levels within the organisation. Provide each group with sticky notes and markers. Instruct groups to generate values that they believe are important to the organisation, one value per sticky note. Encourage participants to think broadly and creatively, focusing on ideals, principles and beliefs that reflect the organisation's ethos and aspirations.

Sharing and Clustering (20 minutes): After the idea generation session, invite each group to share their list of values with the larger group. As values are shared, write them on the whiteboard or flip chart, grouping similar values together. Encourage participants to discuss and clarify the meaning of each value as it is presented. Facilitate a collaborative discussion to identify common themes and patterns emerging from the values generated by different groups.

Refinement and Consensus (20 minutes): As a group, review the clustered list of values and identify any duplicates or overlapping themes. Facilitate a discussion to prioritise and refine the list, focusing on values that resonate most strongly with the organisation's mission, vision and culture. Encourage participants to consider the significance and relevance of each value in guiding organisational behaviour and decision making. Use consensus-building techniques, such as dot voting or ranking, to prioritise the final list of values.

Articulating Values (15 minutes): Once the final list of values has been determined, work collaboratively to articulate each value in clear and concise language. Encourage participants to define what each value means to them personally and how it reflects the organisation's identity and aspirations. Write down the articulated values on a separate sheet or display for all participants to see.

Reflection and Commitment (10 minutes): Conclude the workshop by inviting participants to reflect on the values identified and their significance for the organisation. Encourage participants to consider how they can embody these values in their daily work and interactions. Invite individuals to share their commitment to upholding the organisation's values and fostering a culture that reflects them.

Follow-Up and Implementation (ongoing): After the workshop, compile the list of articulated values and distribute them to all stakeholders. Incorporate the values into organisational communications, policies and practices to ensure they are reflected in all aspects of the organisation's operations. Foster ongoing dialogue and engagement around the values, providing opportunities for reinforcement, celebration and accountability.

Team Activity	*Team/Organisation Purpose Statement*
Objective	This activity is adapted from one called 'Finding Your purpose' attributed to Gustavo Razzetti. The purpose of this exercise is to support organisations or individual teams to devise a purpose statement for why they do what they do and why it matters.
Duration	1–1.5 hours
Materials Needed	Meeting room or designated space conducive to open discussion, Flip chart paper or whiteboard, paper and pens.

Here's how it works:

Introduction (10 minutes): Introduce the purpose of the workshop. Set expectations of open communication, respect and confidentiality. Outline the agenda and any other ground rules.

Split the group into smaller groups of 4–6, the exercise can either focus on team purpose or organisational purpose, don't try to do both at once. How you split the groups will depend on whether the focus of the workshop is Team Purpose or Organisational Purpose.

What does your team/organisation do? (15 minutes): Tell the group to write down in one sentence what their team/the organisation does. Ask them to consider what it delivers, creates or produces. Have them work as a group to collectively agree on a sentence.

If their team/the organisation produces several things, prompt them to find one that represents the majority of the work: for example, *"Our team provides insight data to improve decision making for our leaders"* or *"Our organisation helps companies streamline their finances to maximise profit"*.

Who does your team/organisation work for? (15 minutes): Ask the group to identify the different stakeholders that the team/organisation works for. Again, prompt them to agree on the most important one. For example, *"We support senior executives with critical decision making"* or *"Our organisation helps companies seeking data solutions"*.

What impact does your team/organisation create? (15 minutes): Ask the group to consider what is the impact that their team/organisations hope to create in those that they work for. For example, *"We help our leaders can make quicker decisions, saving time and money"* or *"So our customers can make an informed decision and feel safe"*.

Create A Purpose Statement (15 minutes): Encourage the group to draw those three responses together into on short sentence that defines their team/organisation's purpose in a simple and accessible way. For example; *"Our team processes the latest data and insights to come up with new solutions to ensure our company remains at the forefront of innovation"* Or *"Our organisation provides accessible products so that customers can invest with ease and feel more secure about their financial future"*.

(Continued)

Refine & Debrief (15 minutes): Invite the groups to read out their purpose statements. If you have multiple groups, get them to compare the statements and find points of similarity. As a group refine the different statements until you collectively agree on one that represents the team/organisation. Discuss how you can apply this new purpose statement to your work moving forward and how you could communicate it with the team and the rest of the organisation. Encourage people to print it out, put it on their email signature and agree to review it regularly.

References

Brown, B. (2018). *Dare to lead: Brave work. Tough conversations. Whole hearts.* Random House.

Chouinard, Y. (2016). *Let My People Go Surfing: The Education of a Reluctant Businessman-- Including 10 More Years of Business Unusual* (Revised and Updated). Penguin Books.

Cowan, N. (2010). The magical mystery four: How is working memory capacity limited, and why? *Current Directions in Psychological Science, 19*(1), 51–57.

Schwartz, S. H. (1992). Universals in the content and structure of values: Theoretical advances and empirical tests in 20 countries. *Advances in Experimental Social Psychology, 25*, 1–65.

6 Managing Resilient Teams: Developing Personal and Team Resilience to Support Stress Management and Adaptability

Now here is something you are likely most familiar with: the role of resilience in organisational wellbeing. In this chapter I will break down the construct of resilience, what it is and how we can develop it. Unlike previous chapters this one will not include a selection of activities for you to try out. The reason for this is twofold: the first is that there are currently a huge range of resilience building programmes and activities out there, lots of them free of charge that you can access. The second is that it is my belief that if I can share with you an understanding of what the construct of resilience really is, and the aspects that make it up, then you can be in a position to construct your own resilience interventions tailored to the specific needs of your organisations, your teams and your people. Forgive me for not giving you a simple road map at the end; instead, I'm going to teach you how to build the roads themselves.

The VUCA Age

You may have heard this acronym before. If not then let me welcome you to the VUCA age. It is a period characterised by Volatility, Uncertainty, Complexity and Ambiguity (Johansen, 2012).

This term has its origins in military strategy but translates well into business contexts to describe the changeable, challenging and unpredictable nature of today's world. Let me break down what I mean:

- **Volatility**: Refers to the speed and scale of change in a given situation. Volatile environments are characterised by rapid and unpredictable shifts, making it difficult to anticipate and plan for the future. For a workplace example, imagine a start-up operating in a new and rapidly evolving industry, one where market conditions can change overnight as a result of technological advancements or consumer preferences.
- **Uncertainty**: Describes being unable to predict future outcomes. Unpredictable environments make it challenging to develop long-term strategies. Consider an energy company in an industry facing regulatory changes or policy shifts that could have a significant impact on its operations; however, the changes and their implications remain unclear.

DOI: 10.4324/9781003407577-6

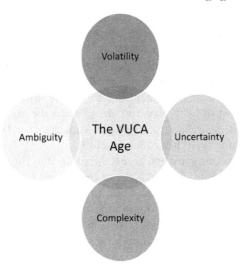

Figure 6.1 The VUCA age.

- **Complexity**: Concerns the high level or intricacy and interconnectedness of systems. Complex environments are ones that involve multiple connected variables and interdependent systems making it challenging to predict exact cause-and-effect relationships or identify straight line solutions. An example of this can be found in companies that manage a global supply chain with multiple suppliers, logistics partners and distributers, each with their own needs and each impacted by their own social, economic and political surroundings.
- **Ambiguity**: Involves a lack of clarity about the meaning or interpretation of information. Ambiguous environments or situations involve a range of possible interpretations or outcomes. It can be hard to make decisions from within the fog of confusion. This can be observed when an organisation attempts to expand into a market influenced by very different cultural norms than those they are used to.

The VUCA lens offers an accurate description of the social, political and professional world many of us find ourselves in today, and the thread that runs across all of it is change. A 'Our relationship to change' is perhaps the best description of the concept of resilience.

Resilience: What Is It?

Well, sometimes it's a buzzword; sometimes it's an excuse to blame individuals for their difficulties rather than organisations taking responsibility for their culture; right now it represents a multi-million-pound global training industry, but in reality, it describes a fundamental aspect of human nature and an important piece in your wellbeing culture puzzle.

The original Latin route of the word *Resilio* means to recoil or bounce back. This is often how it is referred to, as you 'Bounce back ability'. However, this reductionist view ignores the vast amount of research and literature that has focused on defining and understanding this term since the 1800s, gaining significant momentum in the last five decades. In the context of both individuals and teams, resilience refers to the capacity to adapt in the face of adversity and change (Fletcher & Sakar, 2013).

Maybe the first thing to get across in understanding resilience is that human beings don't like change. Let me show you; fold your arms for me. Now you might need to prop this book up, however you are reading it, so you can keep your arms folded as you read. Done? Ok, now with your arms folded I want to remind you that human beings have a natural aversion to change; in the early days of humanity, uncertainty was dangerous. Better to stick with the berries you know than to branch out and try the new berries.

Now, fold your arms the other way for me please. How's that? Feel good, feel nice? No? You don't even like changing the way you fold your arms. You feel discomfort and uncertainty and want to return to the way you are used to, the way you like. This aversion to change is hardwired into your brain. If you practised folding your arms the other way, eventually you would adapt, and I'm sure some of you would adapt faster than others, but initially you have this aversion.

Resilience describes your relationship to change and a mindset which dictates the speed at which you are able to adapt to and tolerate stress. The concept has been explored by researchers in a range of disciplines from ecology to psychology. In an organisational context, it has perhaps been most popularised by the work of Martin Seligman (2011) in his book *Flourish* and in his description of resilience as 'an ordinary magic'. It is just that a power or skill inherent in each of us to different degrees. However, it can be a hard magic to pin down.

There are a range of different models that describe resilience; these are often shaped by the contexts in which they exist. In military settings they include things like mental toughness and physical resilience, and the best one I came across in a military setting was 'grit'. I would like to pause here and ask if you can offer a definition and measurement tool for the concept of grit? No, me neither. In educational settings, you might see things like confidence, composure and commitment, whereas in work settings, you might see robustness, redundancy, rapidity and resourcefulness. You'll notice a real leaning towards alliteration in descriptions of resilience.

I was once tasked, in a military leadership training context, with condensing the last 50 years of research findings on resilience to draw out the definition and trainability of the construct. I will therefore offer you what I found concerning the concept of resilience and the aspects of which can be taught or developed by the organisation or the individual, something relevant

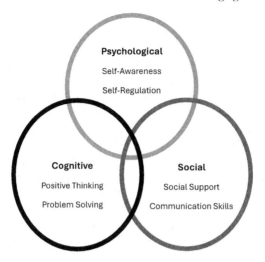

Figure 6.2 Resilience.

to you in cultivating a culture that develops the resilience of your employees and your organisations.

Like a good psychologist, I will separate it into psychological, social and cognitive factors.

Psychological Factors: Psychological resilience describes an individual's internal resources and coping strategies that support them to navigate challenges and setbacks. At its core lies self-awareness and self-regulation.

* **Self-Awareness:** We have covered this skill before, and as you hopefully remember, it describes the ability to recognise and understand your thoughts, feelings, strengths, limitations and motivations. It includes an awareness of your internal state and an understanding of how that internal state may influence your external environment and vice versa. It incorporates aspects of emotional intelligence as well as self-reflection. The more awareness of our internal state in response to adversity or change, the greater our ability to react and respond helpfully. Self-awareness is essential for self-regulation, control of oneself by oneself.
* **Self-Regulation:** This refers to your capacity to regulate your emotions and emotional responses. It involves the use of adaptive coping strategies, such as relaxation techniques, in response to stress, regulating your arousal and remaining 'calm under pressure'. The key component here perhaps is stress management. Much like your reaction to change, your reaction to stress is also hardwired into your brain. Let me give you an example. I'd like you to hold up one of your hands, unclenched and palm facing out; any hand will do. Let's imagine that this is a simple model of your brain. The wrist and the palm represent your brainstem, the earliest bit of your brain;

it regulates things like breathing, heartbeat and circulation and a little bit of fight or flight. Fold your thumb into the centre of your palm. This represents your limbic system, this is the next bit to develop and it wraps around the first bit; you can think of it as emotional responses to your environment. Your thumb nail, that's your amygdala, your brain's smoke detector (more on this later). Next fold your fingers over the top of your thumb to make a fist. This represents your neocortex; all the groovy, 'foldy' bits of your brain that do all the nice logical computer thinking that you probably think you do all the time. Now open up your fingers again. When you get hit with a stressor or a stressful situation, you flip your lid just like this and your lovely rational neocortex becomes disconnected from the rest of your brain and you are just running on fight or flight and emotional responses. To give you an example of that, do you remember at the start of Covid when everyone was bulk buying toilet paper? That is a perfect example of this – logical rational brain disengaged and people just running on fight and flight and fear. You can't eat toilet papers, you can't spend it; it wasn't the cure for Covid, but there we were. Why? Because of this hardwired stress response. Because of your amygdala, when it detects any threat, emotional, social or physical, it goes off. Your self-regulation capacity describes your ability to notice that you have 'flipped your lid', re-engage your logical and rational brain and return to what I describe as a whole-brain position. If you are interested in this concept, I would recommend taking a look at the books *Thinking, Fast and Slow* (Kahneman, 2011) or the more simplistic, but perhaps more digestible *The Chimp Paradox* (Peters, 2013).

Social Factors: The social components of resilience refer to the quality of your interpersonal connections and support networks that support your ability to navigate adversity and stress. This is particularly important when we consider the cultivation of a resilient workforce as it speaks to a collective rather than an individualistic approach. The key aspects of social resilience are your support network and communication skills.

- **Support Network:** The quality of an individual's support network refers to the extent to which they can receive emotional, informational and instrumental support from friends, family and peers in their social network. These connections serve as a buffer against stress and adversity as they supply a sense of belonging, connection, validation and encouragement, serving to bolster our internal resources and provide external coping strategies. Stronger networks with high-quality social connections mean we are able to draw on more resilience. Consider this in the context of the workplace; if I feel connected to and supported by my colleagues and line manager, I am better able to navigate stress and change and find external support when my internal resources are low.

- **Communication Skills:** Drawing from these external social resources however requires a certain level of skill in communication, not only do I need the self-awareness to notice when I am stressed or have flipped my lid, but I need to be able to reach out and communicate that to others. Social and relational intelligence skills come into play here in my ability to express my needs and boundaries and navigate social dynamics and conflict. The greater my communication, the more I am able to enhance my social connections through increased empathy, trust and understanding and draw on these connections to facilitate adaptive coping when needed.

Cognitive Factors: A resilience mindset describes the cognitive process by which I interpret and respond to adversity and change. The key skills required are positive thinking, problem-solving and flexibility.

- **Positive Thinking:** Taken from the world of positive psychology, this process describes an individual's ability to cultivate an optimistic mindset. To understand why this is important, and why it takes work, you need to understand that humans are hardwired to pay attention to negative experiences. We designed to pay attention to and learn from negative information far more than positive information. Like our stress response, this has an evolutionary advantage. The more we pay attention and learn from negative experiences, the more we can avoid them in the future. The problem, however, is that all this practice at noticing negativity changes the way we process information, directing our attention towards when things go wrong rather than when things go right. Positive thinking, outlined in Martin Seligman's concept of 'Learned Optimism', refers to the ability to notice positive experiences in our day to day and refrain negative ones in a more positive light, supporting us to maintain a sense of optimism and hopefulness in the face of adversity. As I say, this form of mental gymnastics takes practice, and I will outline some strategies towards the end of this chapter.
- **Problem-Solving:** This describes an individual's ability to accurately identify challenges, generate appropriate solutions and implement action plans to overcome obstacles. It requires a combination of analytical and creative reasoning and active problem-solving. People who possess this skill understand the need to 'control the controllable'. This speaks to the awareness that in the face of any given situation or 'problem', you have one of three choices. Control it, influence it or accept it. Worrying about it and doing nothing just increases your experience of stress (Covey, 2020). Being able to identify and adapt to problems while maintaining a sense of control facilitates a reduction in stress and an improvement in overall wellbeing.

- **Flexibility**: Cognitive flexibility details an individual's capacity to adjust their thoughts, behaviours and coping strategies in response to new information or changing circumstances. This directly describes your cognitive relationship with change. For those of you who found the arm folding exercise simple and didn't experience any discomfort or confusion, this is a skill you likely have already developed. Cognitive flexibility describes your openness to alternative perspectives, multiple 'truths' and a willingness to explore and accept different approaches. These are the people who seem to be able to tolerate managing multiple high-pressure tasks simultaneously. This skill is important in the workplace when employees are often required to manage dynamic roles and responsibilities. Empathy is often considered an advanced form of cognitive flexibility as it requires you to feel with people and experience their point of view. How are you at lateral thinking puzzles? Good? Well, you probably have a high level of cognitive flexibility. The more cognitive flexibility an individual has, the more able they are to navigate setbacks, learn from them and find novel solutions to problem-solving and opportunities for growth in the face of adversity.

Personal resilience therefore comprises a combination of psychological, social and cognitive resources that support an individual to navigate uncertainty, adversity and change. Each of them, in turn, can be learned; resilience is a skill we can grow by cultivating self-awareness, social support networks, positive thinking, problem-solving skills and cognitive flexibility. These individual resilience factors can be easily adapted to the context of organisational or team resilience.

The foundations of team resilience building can be defined as trust and collaboration, clear communication and shared goals:

1. **Trust and Collaboration:** Resilient teams are able to collaborate, support one another and share responsibility for tasks. They can communicate effectively and openly without fear of reprisal or judgment. They feel a sense of trust and confidence in each other's abilities, decision and intentions. They experience low uncertainty and high psychological safety.
2. **Clear Communication:** Resilient teams engage in open and clear communication channels; they feel able to voice their opinions, ask questions and provide feedback. They are able to share information and seek help when needed, actively identifying challenges and collaborating to overcome them. They engage in active listening and empathy and communicate with clarity. They prioritise collaborative decision making and open dialogue.
3. **Shared Goals:** Resilient teams have a sense of shared purpose through the sharing of explicit goals and vision. They experience a shared sense of purpose, inclusive of shared values, supporting them to maintain motivation when overcoming obstacles as a team. They have shared goals

which establish clear expectations and accountability, aligning effort with objectives encouraging a sense of unity, direction and commitment. They are able to innovate in the face of challenges and effectively identify opportunities for growth.

Workplace Example A: This is what happens when a team lacks resilience:

The customer service team at BounceBack.inc has received a sudden increase in customer complaints following a product recall, and they are struggling to cope. The team became overwhelmed by the volume of complaints and is unable to come up with a collective strategy to manage the crisis. How did this happen?

- **Stress**: *The team felt overwhelmed by the number of customer complaints and struggled to cope with the pressure, with no effective coping strategies; this led to increased stress and burnout.*
- **Lack of Coordination**: *The lack of clear leadership direction and coordination across the team led to confusion and delay in responding to customer complaints. The team struggled to prioritise tasks and allocate resources effectively, and more complaints piled in.*
- **Negative Mindset:** *Instead of viewing the crisis as an opportunity to improve the service, the team became demoralised as they focused instead on the challenges and limitations on their resources instead.*
- **Poor Communication:** *A lack of cohesion and higher levels led to breakdowns in communication across the team, and important information was lost or misinterpreted. The lack of effective communication channels led to increased feelings of disconnection and feeling unsupported.*

Workplace Example B: This is what happens when a team has resilience:

The marketing department at Resilience.inc experienced an unexpected delay on a product launch due to supply chain issues outside of their control; this was a major setback. Despite their initial frustration and disappointment, as well as the pressure both internally and externally, the team was able to work together to support one another and identify creative solutions to maintain their motivation. How did they do it?

- **Adaptability**: *The team quickly adapted to the changing circumstances by exploring alternative strategies and exploring new potential marketing channels to reach their target audience.*
- **Collaboration:** *Team members collaborated closely with other departments, including production and sales, to identify bottlenecks and coordinate efforts.*
- **Positive Mindset:** *The team was able to maintain a positive mindset and focused on finding opportunities for potential growth and innovation. They viewed the setback as a chance to learn and improve.*
- **Effective Communication:** *Open and transparent communication was a key piece of the team's resilience. They kept each other informed of developments, shared updates and ideas freely and sought input from all team members to make informed decisions and promote cohesion.*

Now think about your organisation; which of these situations is more representative of how you respond to crises? Perhaps it is a bit of both. Ask yourself what your organisational resilience looks like. What are your organisation's levels of self-awareness and self-regulation? As an organisation are you able to identify when there are issues and respond appropriately and compassionately from a rational space? Is your organisation characterised by positive relationships between individuals and teams? How do you communicate change and challenges in the organisation? Finally, what is your organisation's relationship with creative problem-solving? Do you focus on the positive or only highlight the negatives? If we want to start developing and nurturing resilience in our workplace cultures, we have to stop thinking of it as simply an individual's ability to 'bounce back' after setbacks and begin to cultivate the skills and environment that support this ordinary magic to grow (Masten, 2001).

Measuring Magic: How to Assess 'Resilience' in Your Organisation

In order to enact strategies to develop resilience, you need to effectively measure it in the first place. Effectively assessing individual and team resilience often requires a multifaceted approach taking into account subjective and objective observations. There are a number of tools that have been designed to do just this.

Assessing Resilience in Individuals

- **Resilience Scales:** Various self-report scales and psychometric questionnaires have been developed to assess individual resilience, exploring factors such as coping strategies, adaptive behaviours and psychological wellbeing (the most popular of these is the Brief Resilience Scale). You can deliver these scales yourself or hire someone in to do it and collect a measure of resilience across your workforce, looking for patterns and provide a baseline benchmark before implementing organisation-wide resilience development strategies or targeted interventions.
- **Behavioural Observations:** Through the use of performance reviews and team assessment, you can gain insights into individuals' resilience behaviours at work. You can observe and make note of how team members navigate challenges, manage stress and collaborate with others to explore resilience dynamics within teams, highlight problem areas and identify strategies for enhancing individual resilience.
- **Structured Interviews:** In addition, you can engage in semi-structured interviews (the performance review is a good example of this) or conduct qualitative surveys (the ones with open text boxes rather than 1–10 scales) to offer opportunities for individuals to reflect on their resilience experiences, strengths and challenges at work. This way you can identify

individual areas of strength as well as areas for development and individual perceptions of personal, team and organisational team resilience to help guide your interventions.

Assessing Resilience in Teams

- **Team Surveys:** You can conduct surveys and questionnaires that assess team dynamics, communication patterns and collaboration to provide data on team resilience. These surveys may include items related to those outlined above, such as trust, collaboration shared goals and shared decision making, exploring resilience strength and areas for improvement.
- **Observational Assessments:** You can directly observe team interactions and behaviours in real time, noting how teams respond to challenges, adapt to changing circumstances and support one another.
- **Network Analysis:** You can explore social networks in the organisation through tools such as Social Network Analysis, also called Sociograms. These techniques can be used to understand a 'community' by mapping the relationships that connect them in a network. These maps can provide information on a team's communication patterns, the flow of information between teams and the quality of social connections. This can help you identify communication bottlenecks as well as resilience strengths and vulnerabilities.

Let me put this into a scenario to show you what I mean. Imagine a software development team at a tech company that is currently undergoing a major organisational restructuring, a common and considerable cause of uncertainty and change in an organisation. The restructuring will lead to significant changes in team composition, individual roles and the assignment of projects. The resultant volatility, uncertainty, complexity and ambiguity have led to increased anxiety and a general drop in wellbeing among the team members, who now have concerns across all levels of Maslow's hierarchy from job security to the future purpose and direction of the team as a whole.
How could you assess the situation to identify resilience and needs?

- **A Resilience and Change Management Workshop:** You could notice that the increase in anxiety and low mood suggests that the team is likely struggling to cope and adapt to the uncertainty. You could organise a team resilience workshop to support them in developing their personal and team resilience and improving their skills in self-regulation and self-care.
- **A Resilience Assessment:** You could conduct a brief assessment or 'pulse survey' to explore their current levels of stress, job satisfaction and perceptions of support. This could help you identify areas for targeted information or greater clarity to directly address their resilience needs.
- **Organise a Fact-Finding Session or Focus Group:** You could arrange a meeting for an open group discussion to invite them to share

their experiences, concerns and suggestions for navigating the current period of changes. You could help them to identify collective strengths and develop action plans for coping with the continuing uncertainty while maintaining a sense of collective purpose.

Understanding that periods of uncertainty and change are the most common source of stress and actively identifying signs of low resilience in your teams can help you provide active interventions and encourage open communication and connection, which, in turn, will bolster their coping strategies and adaptability. In organisations, resilience is an 'Us' problem, not a 'Them' problem.

Improving Individual Resilience

Remember that your organisational system is made up of the individuals who work there; interventions that improve personal resilience will also serve to improve collective resilience. Strategies to develop personal resilience can be aligned directly with the construct of resilience outlined above, including psychological, social and cognitive interventions.

There are seemingly endless resilience programmes out there offering solutions to your organisation's lack of resilience. Some are more effective than others; my suggestion is if you choose to purchase an external product, look at not only the qualifications and experience of who is delivering it but also whether it addresses the directly trainable aspect of resilience as a construct.

In a nutshell, the interventions that have the biggest impact on personal resilience can be divided into stress management techniques (self-awareness and self-regulation), Emotional, Social and Relational Intelligence (ESRI) building strategies (support network and communication) and growth mindset (positive thinking, problem-solving and cognitive flexibility).

Stress Management Training

Resilience relates primarily to our ability to recognise, regulate and adapt to stress. Any strategies we engage in that develop these abilities will, therefore, have a positive impact on our level of resilience.

- **Mindfulness and Relaxation Exercises:** No matter how you do it, activities that focus on developing employees' mindfulness and relaxation strategies will support them with increased self-awareness and self-regulation. Whether its breathwork or meditation class, encouraging regular exercise or offering gym passes, stress management workshops or an inhouse masseuse, consider active strategies to upskill your employees in self-regulation and self-care.
- **Time Management Workshops:** These are another good example of upskilling employees with strategies to reduce stress and feel more in control. Noticing when they are overwhelmed and enacting strategies to manage prioritise tasks will help them in responding to stress and

regulating their system to make better decision during high-pressure situations. Around work hours all can contribute to better time management and reduced stress levels.

Developing ESRI

We covered this concept in more detail in Chapter 1. In essence, it describes our level of self-awareness and self-regulation with regard to our own emotional state, our level of empathy, situational and social awareness and awareness of our and others' needs and relational authenticity. Stress management interventions, like those above, will aid in the development of self-awareness and self-regulation. In addition, if we can provide interventions that develop the remaining skills individually or collectively, we can help develop resilience throughout the system. A number of ESRI interventions are offered at the end of Chapter 1; some other examples include:

- **Cross-Functional Activities:** Any activities that bring your teams together, preferably from across the organisation, will begin to develop their ESRI competencies. The more they can work with, empathise with and learn from each other, the more they will have an opportunity to develop emotional and social skills and the greater their sense of connection and community. These can be team-building days, cross-functional problem-solving activities, collective projects or idea generation sessions, to give a few examples.
- **Personal Presentations:** Organisations often talk about the notion of 'bringing your whole self to work', yet in reality this doesn't often happen. You can build empathy and connection through inviting team members to give short personal presentations to the team. One format I have seen work is having the presentation be made up of 10 photographs of things the individual cares about with 1 minute allowed per photograph. The purpose of the exercise is to encourage team members to share parts of their authentic self, the things that matter to them and motivate them and their values. Team members may feel awkward about such a 'show-and-tell'; however, if we can normalise these activities, we can embed authenticity and openness into our cultures.
- **Talks and Presentations:** Inviting a speaker to speak on a range of topics around their personal stories of resilience, mental health, wellbeing and life experiences can help build empathy and connection, particularly if those speakers are from within the organisation. This action is made even more powerful if these talks are given by members of the senior leadership team. These employees are some of the key culture curators in the organisations and can lead by example with open communication.

Cultivating a Growth Mindset

The 'growth mindset' is a term from the word of positive psychology and refers to our capacity to reframe challenges as opportunities for growth and learning. It is all about problem-solving and cognitive flexibility, encouraging team members to embrace setbacks as valuable learning experiences rather than barriers to growth (Rhew et al., 2018).

- **Encouraging a Learning Culture:** By fostering a culture of curiosity, learning, experimentation and growth, we can help develop a 'learning' rather than a 'blame' culture in our organisation. We can achieve this through regular feedback (outside of simply the performance review). We can make sure to celebrate successes as well as highlighting areas for recovering. Start a meeting with stories of what has gone well rather than always starting it with a 'problem'. Encouraging a learning culture across the organisation, where people feel free to speak up and take calculated risks, can support the organisation as a whole to better respond to challenges and adapt to change (Edmondson, 2012).

Resilience Mindset: Learn to Love 'Discipline'

Engaging in resilience building programmes yourself and providing workshops for staff to attend are effective interventions to develop individual and organisational resilience and improve wellbeing. To embed this and keep it going, however, you will need to maintain commitment to the practices, even when it is difficult. This is where the role of discipline comes in.

I am no discipline expert; most of what I know I gleaned from a workshop run by a colleague and good friend, Major (Retired) Menucha Knebel, titled 'Learn to Love Discipline'. The purpose was to encourage people to change their relationship with the word 'discipline'. It's a word that has negative connotations for many of us as we often associate it with doing something we don't want to do or being 'made' to do something. Menucha instead described discipline as "the highest form of self-love" and the key to a resilience mindset.

She spoke about developing the ability to sacrifice short-term pleasures for long-term growth and rewards, such as improving our mental and physical wellbeing. She described how it is human nature to gravitate towards the path of least resistance and seek comfort wherever possible. Therefore, eschewing comfort in favour of growth is not something that comes naturally to many of us. However, if are able to make a conscious choice to change, such as committing to new practices to build our personal resilience, and demonstrate discipline in that area, we can gradually extend it to other areas of our life.

As we discussed before, change is not easy, and learning to love discipline is a process of training our minds to embrace the discomfort of personal growth. By setting clear goals, acknowledging our progress, being accountable and practising self-care, we create a psychological environment conducive to disciplined action. The combination of resilience and discipline is a winning formula for helping you identify a goal and see it through, and it will serve you well for some of the broader processes of actively designing a sustainable workplace culture, which we will cover later in this book.

Resilience and Culture

Resilience is one of the powerful unseen forces that shapes and is shaped by your organisation's culture. It is the individual, team and organisational ability that will help you successfully navigate change, overcome setbacks, drive innovation, remain disciplined and thrive in these VUCA times. By understanding the core tenets of what resilience is, you can implement simple and accessible strategies to help build it.

Remember, resilience is a team effort; it is not just an individual trait or an individual's responsibility but a collective capacity that runs across the entire organisation. One of the key components is mindset. Resilient organisations demonstrate a growth mindset from top to bottom, championing a learning culture and viewing challenges as opportunities for growth. They are consistent and disciplined in their commitment to growth. Finally building resilience is an active process. It requires measurement, reflection, practice and intervention, in a continuous cycle, to do it right. It also requires support from the leaders and culture curators in the organisation if we want to truly embed it in a sustainable way. The more you can make resilience integral piece of your organisation's culture, the more readily the organisation and its employees will be able to respond to the inevitable uncertainty and change in the VUCA age.

References

Covey, S. R. (2020). *The 7 habits of highly effective people.* Simon & Schuster.

Edmondson, A. C. (2012). *Teaming: How organisations learn, innovate, and compete in the knowledge economy.* John Wiley & Sons.

Fletcher, D., & Sarkar, M. (2013). Psychological resilience: A review and critique of definitions, concepts, and theory. *European Psychologist, 18*(1), 12–23.

Johansen, B. (2012). *Leaders make the future: Ten new leadership skills for an uncertain world* (2nd ed.). San Francisco: Berrett-Koehler Publishers.

Kahneman, D. (2011). *Thinking, Fast and Slow.* Farrar, Straus and Giroux.

Masten, A. S. (2001). Ordinary magic: Resilience processes in development. *American Psychologist, 56*(3), 227–238.

Peters, S. (2013). *The Chimp Paradox: The Mind Management Program to Help You Achieve Success, Confidence, and Happiness.* TarcherPerigee.

Rhew, E., Piro, J. S., Goolkasian, P., & Cosentino, P. (2018). The effects of a growth mindset on self-efficacy and motivation. *Cogent Education, 5*(1), 1492337.

Seligman, M. E. P. (2011). *Flourish: A visionary new understanding of happiness and well-being.* Free Press.

7 How to Expect the Unexpected: Crisis Management and the Value of Pro-Active Wellbeing Strategies

We talked about resilience in the previous chapter; now it makes sense to talk about when we need to draw on it most, in a crisis. In this chapter we will explore what a crisis is, the profound psychological impact of crises on individuals and organisations and how we can prepare for them and effectively respond. In this VUCA (Volatile, Uncertain, Complex and Ambiguous) world, crises are inevitable in any organisation; however, the more conscious effort we put into cultivating a culture of resilience and adaptability, the better we are able to respond and flourish where me might otherwise falter.

What Is a Crisis?

In the modern VUCA world of business, unexpected challenges and crises can arise at any moment, bringing with them threats of organisational instability, loss and failure. We can perhaps define a workplace crisis as any event or situation that disrupts normal operations and poses a threat to the safety, wellbeing, reputation or the successful operation of an organisation. These crises can vary widely in nature and severity, ranging from natural disasters and financial crises to cybersecurity breaches and public relations issues. If I ask you to, I imagine you can bring to mind an experience of crises in an organisation, either the one you currently work in or one you have worked in the past. These experiences are common. Yet despite this, modern organisations are still often 'unprepared', in terms of systems and responses, psychologically and culturally (Boin et al, 2017).

As well as the potential financial or reputational consequences of a crisis, they can have a significant psychological impact on individuals and organisations. Indeed, when you thought of the crises earlier, I imagine the clearest thing you can recall is the experience or stress or worry and perhaps the sense of relief when it was finally over. The psychological impact of crises can be far-reaching, affecting mental health, wellbeing and performance in various ways. These can include:

- **Stress and Anxiety**: Crises are a source of uncertainty. This, in turn, can lead to feelings of stress and anxiety among employees. Uncertainty

DOI: 10.4324/9781003407577-7

and ambiguity, coupled with a pressure to respond quickly and effectively, usually have negative consequences on people's mental health and wellbeing.

- **Fear and Loss:** The potential threat of job loss, financial instability or reputational damage can lead to feelings of fear and insecurity among employees. This lack of safety undermines the very first level of Maslow's hierarchy and can lead to drop in morale and subsequently a reduction in productivity and job satisfaction.
- **Emotional Distress:** We all respond emotionally to crisis situations; this can include sadness, anger, frustration as well as feelings of helplessness and hopelessness, attempting to regulate these emotional experiences while also trying to navigate the demands of the crises and potentially become overwhelming leading to stress and burnout.
- **Leadership Challenges:** During a crisis, leaders and managers may experience increased pressure and responsibility and perceive of increased scrutiny. This can increase their levels of stress and impact their decision making, in turn having a negative impact on the management of the crisis itself.

Mitigation and management of these risks require a certain level of crises preparedness. The more prepared and proactive we are, the more we have a safety net or wellbeing reserve we can call on to navigate the associated uncertainty. Preparing for a crisis will support your organisation in:

- **Minimising Impact:** The more proactive we are, the better we are able to minimise the impact of an unexpected crisis. Rather than wildly bailing out water, we can build a boat with bigger sides. This may take the form of plans, protocols or strategies that support you to act swiftly and reduce potential harm.
- **Maintaining Stability:** Preparation provides a sense of stability and security in times of uncertainty. Knowing that we have strategies and resources, personal and professional, at our disposal can help reduce anxiety and support team members to feel empowered when facing a crisis.
- **Preserving Wellbeing**: Being prepared is not just about having crisis management strategies in place but also about safeguarding organisational wellbeing throughout the experience. Proactive strategies need to focus importantly on employee mental health as much as structural policies.

How to Prepare for an Unexpected Crisis

Cultivating a workplace culture where leaders and staff are able to respond and effectively to crises requires a proactive approach that fosters resilience, open communication, adaptability and a culture of preparedness (Baran & Woznyj, 2020). Some of these may be familiar to you, others are not, but here are the key processes:

1. **Provide Training and Education**
   Offer regular training sessions and workshops on crisis management that include scenario-based simulations and role-playing exercises. Ensure that all employees, at all levels, have the knowledge and skills to identify, report and respond to emergencies effectively.

2. **Develop Clear Protocols, Procedures and Plans**
   Establish clear protocols and procedures for responding to different types of crises. Ensure that these protocols are documented, easily accessible and regularly reviewed and updated as needed. Identify critical systems, processes and resources that are essential for the continued running of your business, and develop redundancy and backup plans to mitigate risks where possible.

3. **Build a Strong Leadership Team**
   Invest in developing strong leadership capabilities at all levels of the organisation. Equip leaders with essential skills for crisis management, including crisis decision making and effective communication and strategies to instil safety and maintain motivation.

4. **Encourage Cross-Functional Collaboration**
   Ensure that your teams are skilled in collaboration across departments. You can do this through cross-functional crisis planning exercises as well as developing a sense of connection across all teams. This will encourage teams to support one another during a crisis.

5. **Promote Flexibility and Adaptability**
   Instil a mindset of flexibility and adaptability among employees and leaders. Provide them skills and tools to support them to adjust strategies in response to changing circumstances.

6. **Provide Support and Resources**
   Plan support services and resources to help employees cope with stress, anxiety and other emotional challenges in advance so that you have them ready in a crisis. This may include access to counselling services, employee assistance programmes or wellness initiatives.

7. **Communicate Transparently and Frequently**
   Communication is key in a crisis, particularly the communication that comes from the top down. Keep employees informed and updated on the latest developments during crises by communicating with transparency and honesty. Accuracy is the antidote to uncertainty.

8. **Learn from Past Experiences**
   The best predictor of future behaviour is past behaviour. Ensure that your crisis planning scenarios include crises that have occurred in the past. Once a crisis is over, conduct post-crisis review to reflect on lessons learned and areas for improvement. Then use these insights to refine crisis response plans, update protocols and enhance future preparedness.

It is strategies like these that will help you effectively transition from being reactive to proactive with regard to crisis management. The more prepared you are, the less of a psychological impact the crisis will have on your organisation as a whole. While these crisis-specific preparedness strategies are essential, some of the skills, competencies and organisational mindsets we covered in the previous chapters will also aid you in expecting the unexpected and navigate the uncertainty of crisis.

Proactive Wellbeing Interventions as a Crisis Preparedness Strategy

From my experience there is a significant link between crisis management capability in an organisation and their engagement with proactive wellbeing strategies. When an organisation prioritises wellbeing as part of their culture, they lay the groundwork for organisational crisis management through enhancing organisational resilience. By engaging proactively in developing the strategies we covered in earlier chapters, you can support crisis management capability on an individual and organisational level.

Firstly, build individual and organisational resilience. This may seem obvious, particular as many of the skills associated with effective crisis management are the same skills inherent in resilience. Support your staff in developing the skills necessary to respond to stress and adapt in the face of adversity. What is a crisis if not a rapid introduction to stress and adversity? The more resilience you can encourage in your staff, leaders and the organisation, the more the system will be able to adapt to uncertainty and change, and the more likely they will come out the other side unscathed or stronger.

Next, improve their decision-making and problem-solving abilities. A key component of the resilience mindset and skill base ensures that you engage your staff team in wellbeing practices to help them reduce stress, enhance cognition and improve decision making. Remember, stressed brains make bad decisions; we cannot access that lovely logical thinking when we are in a state of stress. Therefore, we need to already have learned strategies to regulate our system and bring ourselves back to a whole brain position if you expect people to making conscious, calm decisions. I cannot state empathically enough the impact of stress on cognition. In fact, if you cast your mind back to every questionable decision your organisation has ever made in its history, I would argue they were stress-based decisions.

This leads on well to the importance of equipping your staff team with skills in emotional regulation. Help them develop their self-awareness and emotional awareness so that they have the capacity to notice when they are stressed and do something about it. Their ability to identify and regulate their emotional state will support them to navigate the fear and anxiety associated with a crisis situation and aid them in acting from a position of calm.

Finally, foster a strong sense of cohesion and collaboration across your teams through initiatives, and then focus on cross-functional collaboration, inclusivity and open communication. In doing this you will have constructed teams that are able to collaborate in a crisis, share resources and offer each other mutual support, another foundation stone in personal resilience.

By incorporating proactive wellbeing strategies into your organisation culture, you not only improve employees' overall wellbeing but also equip them with the fundamental tools to respond healthily to a crisis and make effective use of the policies and procedures you have put in place to manage them.

A Step-by-Step Crisis Response Plan

So, we have discussed the importance of preparedness and putting proactive strategies in place, but we still have to navigate a crisis when it occurs.

How to respond will differ from organisation to organisation and crisis to crisis. I have supported organisations through a range of crises over the years, having been around at the start, during and post crises.

But before we start thinking about an action plan, let me first begin by introducing you to a crisis scenario. Welcome to Twitchet Enterprises. It's just a normal day in their busy offices. There are a high-performing company in a high-pressure sector, and like any organisation in a similar space, they run primarily on caffeine. Meeting this need for exotic caffeinated burnt bean water, the company has provided a number of high-end coffee machines in the break room area that are in almost constant use as employees load up on caffeine to keep their motivation and morale high. But what is this? a crisis has emerged! All four of the super deluxe industrial coffee machines have broken down. Right on the busiest of days. Tensions rise, anxiety skyrockets and withdrawal starts to set in.

I will now offer you an overview of a simple five-step crisis plan and we can see how they did.

Step 1: Identification and Assessment

We need to be vigilant to notice when a crisis might be brewing (pun intended). Make sure that you regularly monitor internal and external sources for early warning signs of a potential crisis, such as unusual patterns, emerging risks or adverse events.

When the crisis has been identified, begin by assessing the severity and potential impact, consider factors such as safety risks, impact on wellbeing, operational disruptions, reputational damage and legal implications. Gather all the relevant information together on the nature of the crisis, its causes and potential consequences to help inform decision making. Information is key.

Step 2: Activation of Crises Response

Don't worry, you have trained for this. In your pre-crisis stage, you should have identified a 'Criss Management Team' comprising key leaders and decision makers from across the organisation. Now is the time to activate this team and convene an emergency meeting to assess the situation, review the information and identify and initiate response efforts.

Ensure that your response team includes a dedicated crisis communication team prepared to deliver comms internally and externally for the purpose of updating employees and stakeholders when you have some plans in place.

Now is the time to implement any of your contingency or backup plans from your crisis management plans, adjusting these to meet the needs of the crisis at hand.

Step 3: Communication and Stakeholder Engagement

Kick your comms team into gear and establish clear communication channels and protocols for disseminating information to all stakeholders (employees, customers, suppliers, regulators, etc). Make sure that you provide regular updates and briefings to keep stakeholders informed of developments, actions taken and expectations for their support.

Communication here is key; do your best to ensure that all communication is as transparent, honest and empathetic as possible, acknowledging stakeholder concerns, addressing any questions and demonstrating a commitment to resolving the crisis.

Step 4: Decision Making and Resource Allocation

Utilise your crisis decision-making skills to make evidence-based, timely decisions, taking into account potential risks, benefits and consequences of each option. To aid your decision making and gather the best information available, where possible, consult with subject matter experts.

Identify and prioritise the allocation of resources, personnel and support services to address areas of immediate need and mitigate the impact of the crisis on the wellbeing of employees, the organisation and your customers (in that order).

Step 5: Adaptation and Continuous Improvement

Remember, 'this too shall pass'; no matter how challenging the crisis is, eventually it will end. Do not lose sight of this as you ensure that you monitor the progress of response efforts and be open to adjusting your strategies as you go.

Once the crisis is over, conduct post-crises reviews or evaluate sessions across the organisation to evaluate the effectiveness of the response, identify strengths and areas for improvement and update the crisis management plan accordingly.

Now back to Twitchet Enterprises. In this instance they were caught off-guard. This wasn't something they had prepared for. Recognising the gravity of the situation, the HR department swiftly formed a task force dedicated to resolving the coffee crisis. Comprising coffee aficionados and problem-solving enthusiasts, the task force sprang into action, armed with determination and a shared mission: restore the flow of caffeine to the office (crises identification and task force mobilisation).

As the crisis unfolds, the task force begins to come up with solutions. They reach out to their coffee machine supplier, MagicBeanz, to request their emergency machine repair team while forming cross-functional think tanks to explore possible internal solutions to the caffeine crisis and pool expertise and resources (activation of crisis response).

The crisis comms team, comprising maverick marketeers, come up with catchy slogans to boost morale such as "Caffeine Crisis: We're Brewing Up a Fix!", "Out of Beans, But Not Out of Steam: Hang Tight!" and "No Grounds for Panic: Coffee Relief on the Way!", which they send out to the team alongside progress updates (communication).

After their idea generation sessions, the departments rallied together to find alternative sources of caffeine. Some raided their desk drawers for their emergency packets of Nescafe Instant, while others unearthed long-forgotten tea bags from the depths of the communal kitchen cabinets. With ingenuity and improvisation, they brew makeshift cups of caffeine to continue to fuel the workday and distribute them to the teams most in need of a pick-me-up (decision making and resource allocation).

In the face of adversity, Twitchet employees maintained a positive attitude, injecting humour and levity into the situation. Memes and GIFs featuring coffee-related puns flooded the office Slack channel, eliciting laughter and camaraderie amid the chaos. With a light-hearted approach, employees navigated the crisis with resilience and optimism (resilience mindset).

As the crisis subsides and the aroma of freshly brewed coffee fills the air once more, employees pause to celebrate their triumph. An impromptu coffee-tasting event is organised. Through shared laughter and cups of coffee, the office emerges stronger and more united than ever before. The task force arranges review sessions with teams across the organisation to evaluate what went well, what they could do differently and what plans they could put in place so that nothing this terrible ever happens again (adaptation and continuous improvement).

You see what I mean. No matter the crises, don't panic. Stick to the process, build up your organisation's resilience so that you can rely on it to respond and adapt to the crisis as it occurs, working collaboratively to solve problems. By developing your psychological preparedness, as well as effective policies and procedures, you can build for future crises management, safeguarding your operation and more importantly your culture of resilience and wellbeing.

Cultivating and Readiness Culture

Cultivating a culture of flexibility and readiness is essential for making confident decisions and managing in today's VUCA world. By embracing a growth mindset, practising mindfulness and developing adaptive problem-solving skills, individuals can enhance their ability to navigate crises with confidence and resilience. Do yourself a favour and integrate resilience competencies into your organisational daily practice through proactive preparation, enhancing resilience, strengthening relationships across the organisation and enhancing your organisational culture. By embedding these values into everyday practices and decision-making processes, organisations create an environment where employees feel valued, supported and empowered to thrive, even in the face of crises and uncertainty. This shift in perspective requires re-imagining of our organisational leadership models, moving away from the ego-based leadership of old to contemporary perspectives that encompass the leader's role as a culture curator.

Readiness Culture: Practical Strategies for Developing Crisis Readiness

Alongside the proactive strategies of developing overall wellbeing and resilience across the organisation, I have outlined below some crises readiness activities that you can engage in as an organisation for inspiration. Remember, crises management planning workshops don't always have to be serious.

Team Activity	*Crises Simulation Workshop for Leaders*
Objective	The objective of this activity is to provide leaders with practical experience in identifying and responding to workplace crises effectively. By simulating realistic crisis scenarios, leaders have an opportunity to practice decision making, communication and planning skills in a controlled environment.
Duration	1–1.5 hours
Materials Needed	Meeting room or designated space conducive to open discussion, Flipchart paper or whiteboard, printed crises scenarios (Natural disaster, cyberattack, supply chain failure etc, use both past experiences and imagined scenarios), timer, paper and Pens.

Here's how it works:

Introduction (10 minutes): Introduce the purpose of the workshop. Set expectations of open communication, respect and confidentiality. Outline the agenda and any other ground rules.

Explain the importance of effective crisis management and the role of leaders in leading their teams through challenging situations.

Scenario Briefing (10 minutes): Divide participants into small groups, each assigned to a specific crisis scenario. Distribute printed copies of the crisis scenarios to each group, and allow time for them to review and familiarize themselves with the details.

Simulation Exercise (30 minutes): Start the simulation exercise by announcing the start of the crisis scenario. Set a timer for 30 minutes, during which groups must work together to develop a response plan, assign roles and responsibilities and make decisions on how to address the crisis. Encourage participants to consider various factors, such as safety protocols, communication strategies, resource allocation and stakeholder engagement, in their response.

(Continued)

Group Discussion (20 minutes): After the simulation exercise, reconvene the groups for a debriefing session. Ask each group to share their response plan, highlighting key decisions, challenges encountered and lessons learned. Facilitate a discussion on common themes and best practices in crisis management, drawing insights from the different scenarios.

Reflection & Debrief (10 minutes): Conclude the workshop by summarising key takeaways and learning points from the activity. Encourage participants to reflect on their own leadership style and identify areas for improvement in crisis management skills.

Provide additional resources or reading materials for further learning on crisis management and leadership.

The purpose of the crisis management simulation is to give leaders the opportunity to safely practice and refine their crisis management skills in a supportive learning environment.

Team Activity	*Hollywood Crises Management*
Objective	The objective of this role-play game is to engage participants in a light-hearted and interactive simulation of a crisis scenario. By assuming the roles of crisis responders, 'The Crises Rescue Team', participants will practice their decision-making, communication and teamwork skills in an engaging way.
Duration	1–1.5 hours
Materials Needed	Meeting room or designated space conducive to open discussion, flip chart paper or whiteboard, printed Hollywood crises scenarios (alien invasion, time travel mishap, office prank gone wrong etc), timer, paper and pens.

Here's how it works:

Introduction (10 minutes): Introduce the purpose of the workshop. Set expectations of open communication, respect and confidentiality. Outline the agenda and any other ground rules.

Preparation (15 minutes): Split the group into teams for each crisis scenario. Provide the teams with their scenario sheets, and ask them to assign specific roles to one another (Team Captain, Communications Wizard, Technical Whiz, etc). Invite them to review the scenarios and encourage them to embrace their characters.

Role-Play (60 minutes): Start the role-play game by announcing the start of the crisis scenario and setting a timer for 60 minutes. Participants must work together to respond to the crisis, using their creativity, humour and quick thinking to overcome obstacles and save the day. Encourage participants to stay in character and interact with each other in fun and entertaining ways.

(*Continued*)

Discussion (30 minutes): After the role-play game, facilitate discussion and debriefing session. Explore each character's response to the crisis, highlighting memorable moments, funny anecdotes and lessons learned. Encourage participants to reflect on their communication and teamwork skills, as well as their ability to think creatively under pressure. Discuss common themes and best practices in crisis management, using the light-hearted scenario as a springboard for learning.

Reflection & Learning (15 minutes): Conclude the role-play game by summarising key takeaways and learning points from the experience. Encourage participants to reflect on the importance of teamwork, creativity and humour in crisis management.

Provide additional guidance or resources for further learning on crisis management.

By engaging in light-hearted crises management planning sessions such as The Hollywood Crises Management Game, as well as the more serious ones, we can provide a relaxed approach to practicing crisis management skills. By engaging in even playful scenarios, participants can reflect on their ability to respond effectively to crises while having fun and building camaraderie with their teammates.

References

Baran, B. E., & Woznyj, H. M. (2020). Managing VUCA: The human dynamics of agility. *Organisational Dynamics.*

Boin, A., 't Hart, P., Stern, E., & Sundelius, B. (2017). *The Politics of Crisis Management: Public Leadership Under Pressure* (2nd ed.). Cambridge University Press.

8 Transpersonal Leadership: Leading Beyond the Ego

Embedding the characteristics of a healthy culture such as self-awareness, Emotional, Social and Relational Intelligence (ESRI), motivation, communication, values and resilience requires you to have a conscious and consistent leadership team. To help cultivate and embed a healthy culture, you need leaders with an understanding of the interplay between their own personal experiences and perspectives, the feelings, needs and perspectives of the employee and an awareness of the impact of both on the organisation culture as a whole.

In this chapter we will consider the concept of Transpersonal Leadership, a contemporary leadership strategy, adapted from the world of therapy, that incorporates awareness of oneself, an understanding of 'the other' and an appreciation of the system. It describes a leadership style that encourages leaders to 'lead beyond the ego'. As in previous chapters, I will use tools and real-world examples to highlight the importance of leading with curiosity, compassion and effective communication skills, drawing together ideas from the proceeding chapters to define how we shift from outdated leadership mindsets to a more transpersonal perspective.

Ego-Based Leadership

If we are going to explore the notion of leading beyond the ego, I should probably start by explaining why ego-based leadership, an unhelpful norm, is so unhelpful. When we consider the role of a leader, we often view it as a journey of influence, vision and the shaping of an organisation towards some higher goal. It's a powerful position in many ways. Yet leaders are not robots, they are people, and lurking beneath the surface of most people is an equally powerful force that shapes their attitudes and actions – the ego. When you see this word, you may conjure up Freudian concepts of id, ego and super ego, the subconscious motivations behind our behaviours, the good the bad and the ugly (Maccoby, 2004). You would be right. This is where the concept originated, and each of these aspects of our subconscious can be in a healthy or unhealthy state in relation to the behaviours we engage in. In the context of this chapter and in the context of leaders, we will use a simplified version to

DOI: 10.4324/9781003407577-8

understand the Ego as our internal sense of self-identity and self-importance. The Ego seeks validation, control and recognition. These drives are usually self-focused and may be sought at the expense of collaboration, humility and authenticity.

In yourself, an 'unhealthy' or 'unhelpful' ego state may present in behaviours such as arrogance, refusal to ask for help, refusal to accept failure and denial of mistakes. In leadership ego may manifest in these very same things, as well as micromanagement and authoritarianism. Many unhealthy workplace cultures are filled with ego-driven leaders, either as a result of their own tendencies and understanding of leadership or often as a direct consequence of the unhealthy culture itself.

Ego-driven leaders are driven by the pursuit of power, status, and dominance and prioritise their own interests over the needs of their team or organisation. Their decision making is usually heavily influenced by internal bias, beliefs and a fear of failure, leading to a toxic work culture characterised by fear, mistrust and disengagement. Ego-driven leaders may struggle to empathise with their team and consider their perspectives. This, in turn, results in poor communication, decreased productivity and motivation. It sounds extreme maybe, but think about it. I imagine you have experienced leaders like this, either in past roles or your current one. I imagine that the leader in your mind was prone to favouritism, focused on short-term gains over long-term sustainability, less interested in the wellbeing of employees and the health of the organisation than their own status and position. Ego-based leadership is perhaps best considered as a spectrum of severity; however, at any level we can see how this type of attitude and behaviour can have an impact on organisational wellbeing and culture, from decreased engagement, retention and morale. Let me give you some examples of what ego-based leadership looks like:

Micromanaging Mike: Mike, a department manager, is usually observed constantly hovering over his team members' shoulders, micromanaging every aspect of their work. He insists on being cc'd on every email and requires detailed progress reports multiple times a day. Mike's controlling behaviour creates a tense and stifling work environment, with team members feeling frustrated and demotivated.

Credit-Stealing Carla: Carla, the project manager, is quick to take credit for her team's successes but makes sure to blame others when things go wrong. She frequently undermines her team members' contributions and dismisses their ideas, seeking to maintain control and boost her own ego. Carla's selfish behaviour erodes trust and morale, leading to resentment and turnover within the team.

Status-Checking Steve: Steve, the sales director, is obsessed with his title and position within the company. He constantly seeks recognition and validation from senior leadership and peers, often at the expense of his team members' wellbeing. Steve prioritises his own career advancement over the needs of his team, creating a competitive and cut-throat work environment where collaboration is discouraged.

It's not just employee wellbeing, engagement and workplace culture that are impacted by this approach. Ego-based leadership can be the death of an organisation. Consider the collapse of Enron Corporation. Enron's CEO Jeffrey Skilling, alongside other leaders in the organisation, was characterised by the pursuit of aggressive growth strategies, which eventually led to engaging in fraudulent accounting to artificially inflate the company's stock price. These unchecked egos in pursuit of personal growth over organisational health and employee safety led to a cut-throat culture focused on short-term results and eventually resulted in the 'death' of a billion-dollar company (Peregrine & Elson, 2021).

Another clear example is the story of the Royal Bank of Scotland. This was one of the UK's largest banks, and the ego-based leadership of Sir Fred Goodwin brought it to the brink of collapse. Goodwin's desire for aggressive expansion and market dominance led to them taking significant risks and engaging in 'overpriced' acquisition deals. Goodwin's confidence in his own judgement led him to ignore risk management practices and dismiss concerns from external analysts and experts from within the company about his aggressive strategy. His behaviours cultivated a culture where dissenting opinions were dismissed and eventually led to the bank requiring a bailout from the UK government in an effort to mitigate the consequences a collapse would have on the UK financial system. Goodwin was subsequently fired and eventually stripped of his knighthood (Fraser, 2014).

Maybe these seem like extreme examples, and perhaps you think we have learned better. These cautionary tales don't exist in isolation; similar ego-based leadership disasters or near-disasters can be found in the histories of Uber, Tesco, BP, WeWork, Volkswagen, WireCard and Didi Chuxing, and the list goes on. The ego-based approach to leadership continues to permeate organisational cultures around the world, and it continues to play a significant role in shaping organisational dynamics with significant implications across the system.

Transpersonal Leadership

Contemporary leadership requires more than just the ability to make decisions and delegate tasks. It calls for a deeper understanding of and an engagement in behaviours that develop human connection, a sense of transformative approach that understands these needs, offering a psychologically and systemically informed paradigm shift on leading that prioritises authenticity, compassion and concept of 'service'. Transpersonal leaders move beyond ego-driven tendencies to connect authentically with others, inspire trust and confidence and foster a sense of purpose and meaning in their teams (Dethmer et al., 2015; Mathews, 2021).

More than a simple 'leadership style', such as authoritative, delegating or laissez-faire, Transpersonal Leadership represents a philosophy or mindset emphasising the interconnectedness of the interrelated systems that represent the organisation. Transpersonal leaders Incorporate the skills we highlighted in all of the previous chapters on Emotional, Social and Relational Intelligence (ESRI), resilience, motivation and communication. They are guided by principles of purpose, authenticity and interconnectedness. They consciously seek to cultivate the organisation's culture to prioritise the wellbeing of all stakeholders. The key principles of Transpersonal Leadership can be summarised as:

1. **Authenticity:** Transpersonal leaders lead with integrity and transparency, aligning their behaviours, values and beliefs. They are genuine and sincere in their interactions, inspiring trust and confidence.
2. **Compassion:** Transpersonal leaders demonstrate empathy and compassion towards others, seeking to understand their perspectives and experiences. They prioritise the wellbeing and growth of their team members, fostering inclusive environments characterised by psychological safety.
3. **Service:** Transpersonal leaders lead with a sense of purpose and mission, seeking to make a positive impact on the organisation and world around them. They prioritise the greater good over individual interests, striving to create organisations that serve the needs of all stakeholders.
4. **Interconnectedness:** Transpersonal leaders recognise the interconnectedness of everyone and everything within the organisational system. They understand that they themselves and the people they lead are all part of a larger whole. In recognising this they encourage a sense of unity and collaboration, leveraging their team's diverse skills and experience to achieve shared goals.

In contrast to our micromanaging Mike and friends, meet some transpersonal leaders:

Transpersonal Ted: Ted, the team leader, always puts the needs of his team members first. When one of his team members is struggling with a project deadline, Ted offers to help them consider ideas and provides support every step of the way. He encourages open communication and collaboration, creating a culture where everyone feels valued and supported.
Compassionate Claire: Claire, the CEO of a small company, takes the time to get to know each of her employees personally; if she can better understand their needs, she can better understand how to meet them at work. She organises monthly team-building activities and social events to develop connections and camaraderie among team members. Claire leads with empathy and compassion, creating a positive and inclusive work environment.
Service-Orientated Sam: Sam is the manager of a customer service team. He goes above and beyond to ensure that customers are satisfied and well

taken care of. He empowers his team members to make decisions and take ownership of customer issues, trusting them to do what's best for the customer. Sam leads by example, demonstrating a commitment to service and excellence.

It's not just in the rose-tinted world of our imagination that we can see the benefits of a Transpersonal Leadership perspective. You may be familiar with the outdoor apparel company Patagonia. This organisation is renowned for its commitment to environmental sustainability and social responsibility. CEO and founder Yvon Chouinard embodies Transpersonal Leadership by building a culture of transparency, authenticity and purpose over profit. He encouraged purpose through environmental stewardship, investing in sustainable materials and donating a percentage of profits to grassroots environmental organisations. He prioritised employee wellbeing through generous benefits, flexible working and onsite childcare. Finally, he aligned the company's values with its customers, through transparency and storytelling, fostering a community of people passionate about the outdoors (Chouinard, 2016).

Similarly, Ben Cohen and Jerry Greenfield, co-founders of Ben & Jerry's, embody Transpersonal Leadership through an approach that prioritises social responsibility, employee empowerment and community engagement. They enacted this through integrating a social mission into the core of the business model from the start, committing to using their company as a force for change, advocating for social justice, fair training and environment sustainability. They empower employees by cultivating a culture of inclusivity and creativity, encouraging employees to take part in decision making, advocate for causes they believe in and take part in community service projects. This fosters a sense of purpose and aligns personal values with those of the company (Cohen & Greenfield, 1997).

Finally, The LEGO Group, under the leadership of Niels Christiansen, prioritised social responsibility, environmental stewardship and innovation while fostering a culture of play and learning. LEGO communicates their commitment to sustainability through ambitious goals to reduce their environmental impact and promote responsible manufacturing practices. The company invests in renewable energy and recycling programmes and seeks to minimise waste throughout its supply chain. Christiansen promotes diversity and inclusion within LEGO's workforce through inclusive hiring practices and employee resources groups. They encourage employee collaboration through participatory decision making, cross-functional projects and platforms to share their ideas and solutions to drive business success (LEGO Group, 2022).

Again, these leaders aren't alone; many others are following suit, and similar examples can be found in the inclusive, values-based contemporary leadership strategies of Danone, Unilever, Virgin, The Body Shop, REI, Starbucks and Southwest Airlines. The list isn't as long as the ego-based contemporaries, but it's growing. What seems clear from both theory and practice is that the benefits of adopting a transpersonal approach in leadership extends beyond individual wellbeing to organisational performance and success. By prioritising the long-term sustainability of an organisation's culture, rather than the ego-past short-term wins, leaders can make decisions that benefit not only shareholders but also employees, customers and society. They are able to recognise that success is measured not just by financial metrics but also by the positive impact they have on people's lives and the world around them.

How Transpersonal Are Your Leaders?

Let's take a look at your organisation and reflect on how transpersonal your leaders are. They are a number of measures and strategies to look at this in detail, and I will give some examples of some of these at the end of this chapter. You can ask the following questions of your leaders in the form of a questionnaire if you want, but let's begin with you asking yourself. Reflect on the following questions, answer honestly and rate your responses on a scale of 1–10, 1 being 'not at all' and 10 being 'all the time':

1. *"I prioritise the wellbeing and growth of my team members".*
2. *"I lead with integrity and transparency, aligning my actions with my values".*
3. *"I actively seek to understand the perspectives and experiences of others".*
4. *"I strive to create environments of psychological safety and inclusivity".*
5. *"I am committed to making a positive impact on the world around me".*

How did you do? The higher your score, the more likely you are to be leading with authenticity, empathy, compassion and a service-oriented mindset. What about the organisation? Reflect on the following questions with your organisation in mind:

1. *"Do leaders prioritise social responsibility and ethical decision making in their actions?"*
2. *"Are employees empowered to contribute ideas and solutions to drive positive change?"*
3. *"Does the organisation have a clear mission or purpose beyond profit?"*
4. *"Are there initiatives in place to support employee well-being, diversity and inclusion?"*

Again, note your responses. Is there a difference between how you see yourself and how the organisation is? If so, why do you think that is? You are a part of the system, aren't you?

   If we want to harness the benefits of a Transpersonal Leadership perspective, then we need to reflect on the current approach an organisation is using and where we might be falling short. Two of the simplest methods for beginning this process of self-evaluation are behavioural observations and feedback collection. If you want to understand how your leaders are currently leading, look and listen. Look for Transpersonal Leadership behaviours such as the following.

Empowerment

- How do leaders delegate tasks and responsibilities? Do they trust employees to take ownership and make decisions?
- Do leaders provide opportunities for professional growth and development? Are there mentorship programmes, training sessions or leadership workshops available?
- Do they encourage collaboration and teamwork? Do leaders encourage cross-functional collaboration and value contributions from all team members?

Communication

- How do leaders communicate organisational values and mission? Do they consistently emphasise the importance of social responsibility and ethical behaviour?
- Are your leaders transparent about challenges and setbacks? Do they openly discuss issues and involve employees in problem-solving?
- Do they actively listen to employees and address their concerns? Do they show empathy, respect and understanding in their interactions?

Community Engagement

- What is your leaders' level of external engagement and advocacy? Does the organisation participate in community events, volunteer activities or partnerships with non-profit organisations?
- Do they prioritise corporate social responsibility initiatives? Are there programmes in place to support local communities, environmental conservation or social causes?
- How do your leaders represent the organisation in public forums? Do they advocate for social and environmental issues aligned with the company's values?

Recognition and Appreciation

- Do your leaders show recognition and appreciation for employee contributions? Are there regular acknowledgements, rewards or celebrations of achievements?
- Do they promote a culture of gratitude and appreciation? Do they regularly express thanks and recognition for the efforts and accomplishments of team members?
- How do your leaders respond to feedback and criticism? Do they regularly use feedback as an opportunity for growth and improvement, personally and organisationally?

You can also seek feedback from employees within the organisation as to how they experience leadership, how do they feel they are led. You can engage in a 360-degree feedback process, involving peers, direct reports and key stakeholders. Include questions about their leadership behaviours, communication style and impact. Ask specific questions about leadership practices and organisational values. Encourage respondents to offer constructive feedback to highlight areas of strength and areas for improvement related to Transpersonal Leadership.

By collecting information through reflection, observation and feedback, you can gain valuable insight into the level of Transpersonal Leadership within your organisation as well as identifying areas for improvement and intervention to strengthen these practices.

Strengthening Transpersonal Leadership Practices

Understanding the 'Why' of Transpersonal Leadership is relatively clear when we consider the impact outlined in this chapter and the associated behaviours we covered in the previous ones: promoting a culture of trust and collaboration that nurtures empathy; promoting employee engagement by empowering people to contribute to the organisation in a meaningful way, developing their sense of purpose and satisfaction; and finally, supporting a holistic approach to wellbeing by acknowledging the interconnectedness of everyone in the system and prioritising programmes and practices that focus on physical, emotional and mental health. If you begin to embed the skills outlined in the previous chapters around ESRI, communication, motivation, leveraging meaning and purpose and building resilience, you will already be operating within the transpersonal philosophy. The active strategies for developing a more transpersonal perspective are ones we have covered already: practise active listening, show empathy, lead by example, encourage collaboration, empower staff members, promote a growth mindset and provide proactive support. They apply to both a whole-organisation philosophy and individual leadership actions.

Let me show you: Dr Hayes is a senior leader in a large, dynamic tech firm. She is engaged in a number of high-pressure projects with the purpose of revolutionising their main product. Armed with her leadership experience, she recognises the potential impact that the level of complexity and pressure will have on the team. She is a 'hands-on' leader and aims to lead with a transpersonal mindset, to encourage collaboration and maintain motivation while supporting her team's wellbeing in the long term. She begins by fostering trust. She engages on open discussions with teams, ensuring everyone's voice is heard and concerns are identified and addressed, laying the foundation for honest communication about expectations and needs. She is empathic in her engagement with each team member, appreciating their perspectives, challenges and goals. She actively listens and offers support tailored to their needs. Recognising that diversity of thought is a catalyst for innovation, and something they need for their current programme of projects, she encourages collaboration through cross-functional team projects to explore different viewpoints, engage in creative problem-solving and challenge industry assumptions. By highlighting their impact on the projects beyond the deliverables as well as how they connect to the company's broader mission, she infuses a sense of meaning and purpose. The teams are given relative autonomy, are encouraged to take ownership of their tasks, innovate freely and practise a growth mindset by learning from setbacks and celebrating milestones. Leading by example, Emily ensures that her behaviours align with her values and demonstrates the qualities of resilience, integrity and authenticity that she wants to see in her teams. Throughout it all, Emily prioritises wellbeing and the importance of self-care. She ensures teams take regular breaks, arranges flexible working and proactively provides accessible wellbeing resources so that they are able to sustain their efforts over time and avoid increased stress and burnout.

You can see how a shift in perspective can align all the different qualities and skills that contribute to a healthy work culture and, in turn, develop those same qualities, creating a sustainable system. If only it were that simple. Like everything else we covered so far, a transpersonal mindset takes practices, and you will have to navigate barriers and obstacles to overcoming the ego-based status quo.

Barriers to Adopting a Transpersonal Mindset

The journey towards embracing Transpersonal Leadership can often be met with resistance. These obstacles, stemming from ingrained unhealthy organisational cultures, alongside ego-based beliefs and fears, can hinder the evolution of leaders towards a more compassionate and purpose-driven approach. The first of these are simple misconceptions or scepticism about the value of the approach. Perhaps even after reading this far, you too are sceptical about its value. Hopefully not. Scepticism is a common response to the introduction of 'something new', especially if that something is in direct opposition to the status quo.

Anticipate scepticism and educate your leaders about the key principles and anticipated benefits of Transpersonal Leadership through workshops, training programmes and educational materials (give them this book!). Provide case examples of organisations and bios of transpersonal leaders who have applied this mindset. Draw from the evidence-base to provide real-world examples of the role of Transpersonal Leadership in driving organisational success. One such example is Microsoft. CEO Satya Nadella identified that the status quo at Microsoft was leading to stress and burnout and stifling innovation. Nadella embraced Transpersonal Leadership principles including empathy, inclusion and collaboration to direct a cultural transformation programme. Nadella prioritised culture over strategy to support the company to continue to innovate and thrive in the long term (Nadella et al, 2017).

With this or similar examples, you can make a compelling case for adopting the approach by highlighting the positive knock-on effects on culture, performance and wellbeing. By raising awareness of why and how it works, leaders can gain a deeper understanding of the value of Transpersonal Leadership and its transformative potential. If that doesn't work, then lead by example.

Get buy-in from key senior leaders and executives, and have them embody Transpersonal Leadership principles and highlight the impact. This can provide you with some proof of concept as well as inspire others to the model.

Alongside the anticipated scepticism, there will likely be some psychological barriers to moving away from an ego-based model. One of the most significant of those is fear of vulnerability. For many, leader's vulnerability is still considered as a sign of weakness and a potential threat to their perceived authority and credibility. You will often see this in the façade of invulnerability that many ego-based leaderships continue to project. This attitude prevents the possibility of authentic connection and encourages the same attitude in staff which is disastrous for organisational wellbeing. Next, your leaders may be unwilling to let go of their feelings of power and status in the organisation. Many leaders derive a significant portion of their self-worth from our position, title and sense of authority. They may fear that relinquishing control or sharing decision-making authority will diminish their status or relevance. This attachment to power, coupled with deeply ingrained belief, is usually found in leaders who are generally resistant to change and cling to familiar patterns of behaviour around control and dominance and are reluctant to empower others. Finally, underpinning the other two potential barriers is a lack of self-awareness. Many leaders operate on autopilot: "this is just the way we do things round here", without an awareness of how their actions and words impact others around them. They are usually unaware of the motivations behind their ego-driven behaviours and are often resistant to feedback and self-reflection, anything that might challenge their self-perception. Again, these leaders are the villains of the story; there behaviours and attitudes may be a direct product of the leadership culture they walked into and developed in.

The obstacles to adopting a transpersonal perspective are not insurmountable, however. To overcome these and cultivate a workplace ecosystem that can support a transition to a more transpersonal mist, you can actively cultivate self-awareness. Using some of the practices outlined in the previous chapters, you can encourage leaders to build self-awareness through regular feedback and self-reflection to identify their ego-driven behaviours and their function. You can also create a workplace environment where it is safe to practise vulnerability. Some of the most effective programmes for this are when leaders are encouraged to actively share their own challenges, failures and coping strategies openly within the organisation. One example I can bring to mind is the multinational professional services firm PwC. It is common knowledge that they produce videos of leaders from across the organisation sharing personal stories of wellbeing and their strategies for managing stress. This behaviour serves to model vulnerability and resilience for all employees, not just leaders. For the more stubborn leaders, you can offer coaching and mentoring programmes to help identify barriers to a shift in leadership perspective, provide guidance and feedback as well as encouragement and opportunities for growth. You can encourage them to see the value of delegation and empowering their staff team's while coaching them to identify their personal blind spots and the limiting beliefs that are getting in the way of change.

Transpersonal Leadership and Culture

Overcoming obstacles and resistance to moving beyond ego-based leadership requires courage and commitment to personal and organisational growth. By addressing scepticism, cultivating self-awareness, encouraging vulnerability and promoting embracing change leadership can begin the shift towards a transpersonal model of leadership and cultivating the organisational culture that they want and moreover one that employees need to continue to grow in the long term. Develop a more compassionate, focused and purpose-driven approach to leadership and reap the cultural benefits, including improved alignment, engagement and the most talked about of cultural needs, psychological safety.

Transcending the Ego: Practical Strategies for Developing a Transpersonal Leadership Perspective

As well as engaging in the strategies outlined in previous chapters for developing empathy, self-awareness, self-regulation and problem-solving, you can actively engage in an assessment of your current transpersonal capabilities through self-reflection, questionnaires and 360-degree feedback as highlighted in this chapter. In addition, you can actively cultivate a transpersonal perspective by using some of the workshop activities outlined below.

Team Activity	*Leading With Purpose*
Objective	This workshop activity provides a structured framework for leaders to explore and connect with their values and purpose, fostering self-awareness and empowering them to lead with authenticity integrating these skills into a transpersonal approach.
Duration	1–1.5 hours
Materials Needed	Meeting room or designated space conducive to open discussion, paper and pens.

Here's how it works:

Introduction (10 minutes): Introduce the purpose of the workshop. Set expectations of open communication, respect and confidentiality. Outline the agenda and any other ground rules. Provide a brief overview of Transpersonal Leadership, emphasising the importance of values, purpose and empathy in leadership.

Values Identification (15 minutes): Ask participants to take a few minutes to reflect on their core values as leaders. Distribute sheets of paper and ask participants to write down their top three values that guide their leadership approach. Once everyone has identified their values, invite a few volunteers to share their values with the group and briefly explain why they are important to them.

Purpose Reflection (20 minutes): Facilitate a guided reflection on purpose by asking participants the following questions:

(Continued)

- *What motivates you as a leader? What drives you to lead?*
- *What impact do you hope to make within your organisation and beyond?*
- *How do your personal values align with your leadership purpose?*

Encourage participants to write down their reflections.

Group Reflection & Discussion (15 minutes): Divide participants into small groups (3–4 people per group). In their groups, invite participants to share their reflections on values and purpose with each other. Facilitate a discussion within each group, encouraging participants to explore common themes, share insights and offer support and feedback to each other.

Action Planning & Close (15 minutes): Ask participants to individually reflect and write down one specific action they can take to integrate their values and purpose into their leadership practice. Encourage participants to consider how they can demonstrate empathy, foster collaboration and promote well-being within their teams and organisation.

Afterwards bring the group back together, and invite a few volunteers to share their action plans with the larger group. Conclude the workshop by emphasising the importance of aligning values and purpose with leadership actions to create positive change and inspire others.

Team Activity	Empathy In Action
Objective	To cultivate empathy, vulnerability and deepen understanding among leaders, fostering a culture of compassion and connection within the organisation.
Duration	1 hour
Materials Needed	Meeting room or designated space conducive to open discussion, flip chart paper, markers, sticky notes, paper and pens.

Here's how it works:

Introduction (10 minutes): Introduce the purpose of the workshop. Set expectations of open communication, respect and confidentiality. Outline the agenda and any other ground rules. Discuss the importance of empathy in building trust, fostering collaboration and promoting well-being within teams and organisations.

Empathy Mapping (20 minutes): Divide participants into pairs. Provide each pair with a piece of flipchart paper and markers. Explain that one partner will share a challenging experience they've had at work, while the other partner will act as an empathetic listener. Encourage the listener to actively listen, ask open-ended questions and express empathy through body language and verbal cues. After 10 minutes, switch roles so that each participant has the opportunity to share and listen.

Reflection & Discussion (15 minutes): Bring the group back together, and invite participants to reflect on their experiences as both speakers and listeners. Facilitate a discussion on the importance of empathy in leadership, exploring how empathetic listening can enhance communication, build trust and strengthen relationships within teams. Encourage participants to share insights, challenges and strategies for incorporating empathy into their leadership practice.

(Continued)

Empathy Action Planning (15 minutes): Distribute sticky notes to participants, and ask them to individually reflect on one action they can take to demonstrate empathy in their leadership roles. Encourage participants to consider specific behaviours or practices they can implement to show empathy towards their team members. Invite participants to share their action plans with the group, either verbally or by posting their sticky notes on a wall or whiteboard.

Reflections & Close (10 minutes): Conclude the workshop by emphasising the importance of empathy as a cornerstone of Transpersonal Leadership. Encourage participants to commit to their empathy action plans, and support each other in practicing empathy in their daily interactions.

Team Activity	Peer Coaching Circles
Objective	Peer coaching circles support leaders to connect with their peers to share insights, challenges Transpersonal Leadership. They represent structured and supportive environment where participants can engage in collaborative learning, reflection and peer support, all key aspects of a Transpersonal approach. Facilitating a peer coaching circle exercise requires effective facilitation skills, active listening, and a commitment to creating a supportive and empowering learning environment. By providing structure, guidance and opportunities for reflection, facilitators can help participants develop their coaching abilities, gain new perspectives and enhance their professional growth and development.
Duration	1.5–2 hours
Materials Needed	Meeting room or designated space conducive to open discussion.

Here's how it works:

Ground Rules (10 minutes): Welcome participants and define the purpose and objective of the exercise including professional development, skill-building or problem-solving. Identify specific objectives for the session, such as exploring challenges, sharing insights or providing feedback. Establish group norms of active listening, open communication, constructive feedback, respect and confidentiality.

Coaching Circles (60 minutes): Divide participants into small groups of 4–6 people, ideally with diverse backgrounds and perspectives. Explain the coaching process, including the roles of the coach and coachee. The coachee shares a specific challenge, goal or issue they are facing, providing context and relevant background information. The coaches then ask probing questions to help the coachee gain clarity, explore potential solutions, and identify action steps. Encourage coaches to provide supportive feedback, share relevant experiences and offer perspectives that may help the coachee gain new insights. Emphasise the importance of active listening, asking powerful questions and providing non-judgmental feedback.

(*Continued*)

Allocate equal time for each participant to share and receive feedback. Designate one participant as the coachee for each round, while the remaining participants act as coaches. Rotate the roles of coachee and coach for each session to ensure everyone has an opportunity to receive and provide feedback.

Reflective Discussion (20 minutes): After everyone has engaged in a coaching session, facilitate a reflective discussion where participants share observations, insights and lessons learned. Encourage participants to reflect on their coaching experience, including what worked well, what could be improved and what they learned about themselves and others.

Action Planning (10 minute): Encourage participants to turn their insights from the peer coaching circle exercise into actionaable steps for their professional practice. Schedule follow-up sessions or check-ins to assess progress, provide ongoing support and reinforce learning.

Summarise & Close (10 minutes): Summarise key takeaways and insights from the peer coaching circle exercise. Thank participants for their engagement and contributions, and encourage them to continue supporting each other beyond the session.

References

Chouinard, Y. (2016). *Let my people go surfing: The education of a reluctant businessman – including 10 more years of business as usual (Revised and updated)*. Penguin Books.

Cohen, B., & Greenfield, J. (1997). *Ben & Jerry's double-dip: Lead with your values and make money, too*. Simon & Schuster.

Dethmer, J., Chapman, D., & Klemp, K. W. (2015). *The 15 commitments of conscious leadership: A new paradigm for sustainable success*. Dethmer, Chapman & Klemp.

Fraser, I. (2014). *Shredded: Inside RBS, the bank that Broke Britain*. Birlinn.

LEGO Group. (2022). Sustainability progress report 2022. [Online] Available at: lego.com/cdn/cs/aboutus/assets/blt767f97833484d573/LEGO_Sustainability_Progress_Report2022_FINAL.pdf (Accessed: 01 March 2024).

Maccoby, M. (2004). Narcissistic leaders: The incredible pros, the inevitable cons. *Harvard Business Review*, 82(1), 92–101.

Mathews, J. (2021). Spiritual theory of leadership effectiveness. *IUP Journal of Organisational Behavior*, 20(1), 52–74.

Nadella, S., Shaw, G., & Nichols, J. T. (2017). *Hit Refresh: The Quest to Rediscover Microsoft's Soul and Imagine a Better Future for Everyone*. HarperBusiness.

Peregrine, M., & Elson, C. (2021). Twenty years later: The lasting lessons of Enron. *Harvard Law School Forum on Corporate Governance*, 5 April. Available at: corpgov.law.harvard.edu/2021/04/05/twenty-years-later-the-lasting-lessons-of-enron/ (Accessed: 01 March 2024).

9 Psychological Safety: Enhancing a Culture of Belonging

I've mentioned this concept a couple of times in this book so far. This is another of those workplace culture concepts you have likely heard about, particularly in recent years. In this chapter I will offer an overview of what we are talking about when we are discussing the value of psychological safety at work and how you can go about building it in a step-by-step way.

Psychological Safety: What Is It?

Firstly, have you ever worked somewhere where you felt like you were walking on eggshells? Where people were afraid to share a new idea or alternate solution to a problem or afraid to admit when they didn't understand something? These are examples of workplaces that lack psychological safety.

Psychological safety is about feeling able to take risks and show vulnerability in front of others, in a personal or professional environment, without fear of embarrassment, rejection or punishment. It's one of the most overlooked but impactful aspects of what makes a workplace culture successful. At its core it describes the degree to which individuals within a group feel safe to take interpersonal risks, voice their opinion, challenge, innovate and express themselves.

The concept of psychological safety was born out of research into organisational behaviour and management theory looking at the interplay between leadership style, culture and engagement in a workplace. The phrase was coined in the 1990s by Harvard Professor Amy Edmondson and has continued to grow in popularity. In brief, her original research was examining the impact of team 'connectedness' in medical teams on the number of 'errors' in their work. She expected to find that teams that were more connected had less errors. Instead, what she found was the more connected the team was, the more errors were reported. What was happening was that the less connected teams, the teams with lower psychological safety, had more errors but were more likely to cover them up or not say that something had gone wrong. You can see the issue here. A rigid culture, where people are afraid to speak up stifle's creativity, leads to errors being missed as well as opportunities for

DOI: 10.4324/9781003407577-9

growth and erodes trust, all of which are unhelpful in an age of rapid change (Edmondson, 2019).

Have a think about your organisation. Ask yourself the following questions to reflect on the current level of psychological safety:

1. *Do I feel comfortable speaking up and voicing concerns or differing opinions in meetings?*
2. *If I make a mistake at work, do I feel able to openly acknowledge and discuss it without excessive fear of repercussions?*
3. *Are people's ideas, inputs and suggestions genuinely listened to and considered, even if they come from more junior members of the team?*
4. *Do I feel safe to ask questions when I'm uncertain about something or need clarification without worrying about looking stupid or incompetent?*
5. *Do failures tend to be treated as sackable offenses, or are they viewed as learning opportunities to reflect and improve?*
6. *Are people able to candidly and constructively call out problems or issues they see without fear of retaliation or being labelled as negative?*
7. *Does the leadership model the behaviours they want to see in terms of openness, humility and not being defensive when receiving feedback?*

If you can mostly answer 'yes' to these questions, it may suggest you are working in an environment of psychological safety, bravo. If you have more 'No's' or 'it depends', it may suggest that there is some work to do.

Psychological safety is more than just a buzzword or a contemporary trend. It is a key enabler for employee wellbeing and organisational success. Research and real-world examples have consistently demonstrated the impact of psychological safety on team performance, resilience, creativity and adaptability. In addition, it is directly linked to employee job satisfaction, lower levels of stress and burnout and greater staff retention, which makes sense: if we don't feel safe, we aren't going to stick around for long. Developing a culture of psychological safety seems obvious, no? It is easier said than done. Our innate fear of rejection by the workplace tribe and the nature of ingrained cultures and unhelpful leadership style act as barriers to adopting this far healthier approach. However, if we can understand how and why it works and explore simple strategies to begin to embed it, we can start dismantling some of the barriers to safety, innovation and authentic collaboration.

Real-World Examples

The shift towards greater psychological safety is becoming a tried and tested method for developing a healthy workplace culture at a number of large international organisations. Take, for example, Project Aristotle, a piece of work conducted at Google across 180 teams, to try and understand why some of their teams were excellent collaborators and others less so. After collecting a large amount of data including traditional factors associated with performance such as shared vision, clear roles, team size and work-life balance, they found the single most important factor for team success was psychological safety. They defined this as employees feeling that they could engage in 'conversational turn-taking' and can be candid with teammates without getting rejected. They identified that those teams with higher psychological safety fostered an environment where everyone felt able to take risks, voice half-baked ideas, admit mistakes openly and question the group's direction without fear of embarrassment or negative repercussions. This created an atmosphere of mutual trust and respect. In contrast, teams with low psychological safety reported feeling guarded, risk-averse and resistant to offering up ideas that might get shot down. This stifled open communication and hampered creativity. Google went on to use these findings to shape their people management strategies and employee training to cultivate a culture of safety and connection (Duhigg, 2016).

A similar focus on psychological safety can be found at Netflix with CEP Reed Hasting's portion of what they describe as a 'Culture of Candour', which encourages radical transparency and a culture of 'freedom and responsibility' when team members are encouraged to offer candid feedback and engage in constructive debate about companywide decision and strategy. The animation giant Pixar has also followed suit by encouraging a 'feedback culture' where employees are encouraged to offer ideas and feedback, even with regard to leadership behaviours, without fear of reprisal. Pixar president Ed Catmull rightly appreciates that a culture of openness is one that allows team members to innovate and improve until they achieve excellence (Catmull & Wallace, 2014; Hastings & Meyer, 2020).

Building Psychological Safety: How Do You Do It?

Before we get into how we can start to build, it lets first take a look at a workplace sorely lacking in the Psychological Safety department. Picture 'The-WayItIs.inc', a bureaucracy of matching white shirts with different coloured collars and cuffs, where conformity is prized and dissent is met with disdain. The atmosphere is stifling, and the environment is one characterised by hushed conversations, wary glances and meetings that take the form of solemn rituals where new ideas are met with scepticism or ignored. Sarah, an enthusiastic marketing coordinator, sits in the weekly team meeting with a bright new idea for a social media campaign for their latest product. Team lead Roger is running the show however, and he's is not a man to have his flow interrupted, especially with something he does not consider 'worthwhile'. He has a reputation for shooting down new ideas with an eye-roll or dismissive comment. Sarah is conscious that the team has been in a creative rut for a couple of months now and she thinks she is really on to something with this campaign, although at the thought of sharing it, her hands become clammy, and she is picturing Roger dismissing her idea out of hand and the resulting smirks from the rest of the team. She sees other colleagues flinch or flounder when Roger asks for updates on their projects or describes his disappointment at this quarter's numbers. Speaking with her colleague after the meeting, she says:

> *It's like a minefield in there. I feel like we are damned either way, if we keep going as we are, we'll get shouted at for not meeting targets and if we offer an alternate solution, we'll get shut down for not doing as we're told!.*

And there it is – the fatal dynamic in this office. A culture where leaders and dominant personalities use public shame as a tool to silence dissent and maintain motivation to pursue the perceived 'right path' – an environment without psychological safety, where offering up genuine thoughts and taking interpersonal risks is too terrifying to attempt. The suppression of candour and idea sharing means the team's work is suffering. This is a workplace where people endure rather than thrive.

The essential elements that we need to encourage to develop a culture of psychological safety are open communication, trust and cohesion and adaptability. Firstly, communication is the cornerstone of psychological safety; just as in Professor Edmondson's research, it is the ability to speak up, sharing ideas and identifying issues. Strategies that promote vulnerability, candour and healthy discourse go a long way in boosting psychological safety over time (Edmondson, 2019).

We want to promote an atmosphere where teams feel comfortable sharing their thoughts, ideas and concerns without fear of judgement or reprisal. You can do this through encouraging active listening, empathy and constructive feedback, qualities we explored in previous chapters. In addition, we can celebrate diversity of thought, encouraging cross-functional teamwork and the sharing of diverse perspectives. In an age of cancel culture, we may sometimes lose sight of the fact that differing viewpoints can lead to richer discussions, innovative solutions and better outcomes. Find ways to encourage team members to challenge assumptions, think creatively and consider alternative viewpoints when problem-solving. The first step to doing this is leading by example. How is your communication? What about your leaders? Leaders serve as role models when it comes to demonstrating psychological safety at work. Embodying values of vulnerability, authenticity and empathy inspires trust and safety. How leaders choose to communicate will directly influence how employees express themselves or not.

Whether psychological safety is high or not is often best observed in the dynamics of the team meeting. Here, surrounded by our peers and organisational hierarchy is where feelings of shame, fear and uncertainty are most apparent. In an unsafe environment, your amygdala is on high alert looking out for any potential psychological or social threat. This is also a perfect environment to begin to model psychological safety and embed practices that encourage it. One simple strategy is to assign roles to group members in the meeting that promote qualities of psychological safety and give permission to speak up. My favourites include 'The Elephant Spotter' – whose job it is to spot the elephant in the room and highlight the things that are not being discussed but are influencing the 'problem' or the conversation. Next is the 'Rabbit Catcher' – whose job is to notice when the group has gone down a rabbit hole and the conversation is veering off course or getting bogged down in unhelpful discussion. Finally, 'The Giraffe' – whose role is to stick their neck out and put forward different opinions, or counter arguments to the topic of discussion, to avoid groupthink and encourage creative thinking. These roles can be assigned actively at the start of each meeting and rotate between employees for subsequent meetings. They are a method to give permission to take risk, challenge the narrative and keep the group accountable.

Another favourite of mine, and one I have deployed with success in a number of organisations, is to encourage the establishment of the weekly 'Challenge Session'. This can be incorporated into existing weekly meetings or exist as a stand-alone 30-minute session. This meeting is an opportunity for employees across the organisation to present something they've been struggling with or an area where they need advice. Employees from across the organisation are invited to attend, particularly those working in different areas, and offer their advice and support. This process normalises highlighting problems, encouraging cross-functional teamwork and seeking diverse perspectives when it comes to problem-solving.

Trust and cohesion are foundational elements for fostering psychological safety and cultivating an environment where people feel secure enough to take interpersonal risks. Without a baseline of trust and team cohesion, people will be hesitant to make themselves vulnerable by speaking up, admitting mistakes or offering unconventional ideas. Trust in this instance refers to team members having the confidence that their intentions and efforts won't be misconstrued or used against them in a negative way. It's about believing your colleagues have your back and are operating with goodwill, holding that unconditional positive regard we spoke about previously. Cohesion refers to the bonds, camaraderie and shared commitment that unite a team together. High cohesion means people feel like they're part of a supportive in-group, with a shared sense of goals, values and direction. When trust and cohesion are lacking, employees will instinctively go into self-protection mode. They will hold back perspectives and censor themselves to avoid rocking the boat. Innovation and constructive disagreement get stifled in service of not disrupting the status quo. However, when trust and cohesion flourish, people feel safe enough to engage candidly in highlighting uncertainty, taking risks and sharing ideas, qualities essential for teams and organisations to thrive.

Developing greater trust and cohesion can be done though actively encouraging interpersonal risk-taking. Provide opportunities for team members to take small, repeated risks around being vulnerable with each other such as personal presentations or 'Challenge Sessions'. Consistency in this regard builds trust. You can find ways to encourage greater social connection in your team through social events, informal gatherings and offsite team-building activities, providing opportunities for colleagues to connect on a personal level. Engaging staff in fun activities outside work allows them to bond over shared experiences and present themselves authentically to one another. In the vein of authenticity, you can champion individuality and diversity through creating forums for people to share aspects of their unique background, interests and perspectives. Build and encourage participation in special interest groups at work in line with your organisation's Equality, Diversity & Inclusion. You can organise talks from in-house speakers in line with this also. Encourage self-disclosure and get people used to opening up on a personal level by having them share core values, life experiences or what motivates them. This is best initiated through modelling vulnerability as a leader, having a leader share their personal values, struggles and strategies to the whole organisation. Something like this can also help you encourage shared values, goals and norms to guide team behaviour and decision making. You can encourage team members to actively participate in defining these shared principles, fostering a sense of ownership and accountability. At Google, for example, teams are encouraged to create 'team contracts' that outline expectations for behaviour and interaction.

Finally, enhanced adaptability is crucial not only for resilience, as we have seen previously, but also for fostering psychological safety. A culture of adaptability is one where people feel comfortable taking risks, experimenting and learning from failures or mistakes. Adaptability allows organisations and teams to pivot quickly, embrace change and continuously develop their processes and strategies. In psychologically unsafe environments, people tend to cling to the status quo, fearing that any change will be met with harsh consequences. This rigidity gets in the way of innovation, creativity and growth. However, when adaptability is valued and encouraged, people feel empowered to suggest new ideas, challenge existing assumptions and take calculated risks without the fear of being punished. Adaptable organisations understand that the 'only constant in life is change' and choose to develop a mindset of curiosity within a culture of learning. In this culture employees feel safe to voice their concerns, share their perspectives and offer solutions. We can build such a culture through highlighting the importance of learning. Provide opportunities for employees to attend training programmes, workshops or conferences that expose them to new ideas and perspectives and actively encourage them to attend. Embrace experimentation through dedicated spaces or initiatives for employees to test out new ideas like 'Challenge Sessions' or cross-functional idea workshops. Make sure to celebrate 'failed experiments' as learning opportunities, and encourage employees to reflect as a group and share their insights. The latter only works when we encourage cross-functional collaboration and regularly expose employees to different perspectives, processes and challenges.

The fear of failure is one of the biggest barriers to adaptability; however if you can cultivate a growth mindset where people learn to embrace challenges, persevere through setbacks and see the opportunity in the error, you can overcome it. However, the leaders will need to model this behaviour if it is ever going to permeate across the whole system. Leaders and the organisation as a whole need to be open to feedback, willing to adjust their approaches and transparent about their own learning. Be an organisation that celebrates positive risk-taking rather than shying away from it. Tesla founder Elon Musk offers a real-world example of this in the decision to vertically integrate and build their own battery manufacturing facility for their electric cars, the Gigafactory. This was a huge undertaking and a represented significant financial risk, with Tesla investing billions into a technology that was still largely unproven at a mass scale. However, Musk's willingness to take this leap created a culture where employees felt empowered to think big and challenge conventional wisdom. Employees felt that even if their ideas seemed far-fetched, they would be given a fair hearing and the opportunity to make their case. This psychological safety to voice daring concepts without fear of immediate dismissal or ridicule fostered an innovation mindset. During the factory's' construction, Musk encouraged employees at all levels to identify potential bottlenecks and surface concerns. He made it clear that he valued candour over blind loyalty, urging people to "apply rigorous analysis" even if it meant delivering hard truths he didn't want to hear. This openness to scrutiny signalled it was safe to take the risk of raising issues. Musk also modelled vulnerability, admitting mistakes and changing course when warranted. For example, after years of resisting artificial intelligence's (AI) assistance for manufacturing due to excessive optimism about their capabilities, he eventually acknowledged AI's importance and rapidly invested in implementing it. This willingness to be wrong created psychological safety for others. The outcome was a workforce united in pushing boundaries and embracing intelligent risks (Vance, 2015).

By celebrating audacious risk-taking and making it okay to voice half-baked ideas, question assumptions and course correct, Musk cultivated an environment of psychological safety that unlocks his team's full innovative potential. You don't have to go this big to start developing a psychologically safe culture, however. Small, consistent actions to promote vulnerability, candour and open communication can go a long way in boosting psychological safety over time. An open culture unlocks significant benefits for teams and organisations.

Common Barriers to Developing Psychological Safety at Work

As we have established in previous chapters, change takes time, and we need to acknowledge potential barriers to doing things differently. Encouraging greater psychological safety is no different. First and foremost is overcoming people's innate fear of speaking up, voicing their opinion and going against the grain. This hesitation to ask questions or share concerns stems from deep-rooted human instincts and experiences that can be difficult to overcome in professional settings. At its core, the fear of speaking up is driven by ancient survival strategies to help us avoid being ostracised from the tribe. As such our amygdala is hardwired to avoid potential social rejection or embarrassment at all costs. When we consider speaking up at work, the amygdala sounds the alarms about the potential social threat, and this results in instinctive feelings like anxiety and fear. We want to avoid looing stupid, damaging social relationships and potential reprisals or negative consequences. Our past experiences of speaking up and being shut down, or witnessing it in others, may reinforce this fear increasing the perceived risk. Look out for comments in meetings like "Sorry, I may be missing something here, but..." or "This is just my perspective, but..." as these highlight the fear people may be experiencing about overstepping or putting themselves 'out there'. Overcoming these deeply ingrained fears of speaking up requires building a foundation of strong psychological safety. Simply recognising the human roots of this hesitation is an important first step.

The second biggest barrier is a systemic lack of trust in leadership. Trust is fundamental to psychological safety, and when employees lack trust in their leaders, it impacts their confidence and prohibits open communication. This lack of trust may arise from perceptions of leaders as untrustworthy, inconsistent or lacking integrity. The Enron scandal we discussed earlier, for example, serves as a clear example of how unethical leadership practices can lead to a culture of fear and silence. Beyond the perception of leaders as unethical, if employees perceive their leaders as punitive or vindictive, they may be hesitant to voice concerns out of fear of retaliation. In addition, when there is a lack of clarity around decision-making processes, organisational changes or performance metrics, it can breed suspicion and mistrust among employees. Also, when leaders are perceived as playing favourites or exhibiting biases (whether conscious or unconscious), it can lead employees to believe that speaking up or taking risks will be met with unequal treatment or unfair consequences. Overcoming this challenge requires leaders to demonstrate authenticity, transparency, open communication, equality and accountability in their actions and decisions. When employees trust that their leaders have their best interests in mind and will support them in taking calculated risks, they are more likely to feel psychologically safe and willing to fully engage.

Building psychological safety is an ongoing journey that requires commitment, effort and continuous improvement from leaders and teams. By addressing common challenges and implementing practical strategies for fostering trust and collaboration, organisations can create environments where employees feel safe, connected and empowered to contribute fully. Through ongoing measurement and evaluation, organisations can also track progress, identify areas for improvement and ensure that psychological safety remains a priority.

Promoting Psychological Safety: A Step-by-Step Guide

If you want to design a strategy to increase psychological safety at work, make it clear and accessible.

Step 1: Leadership Commitment

Start at the top, and ensure that leaders at all levels of the organisation understand the benefits of psychological safety and are committed to developing it through their words and actions. Encourage them to lead by example and model vulnerability, openness and empathy in their interactions with employees. Highlight the necessity for clearly communicating the importance of psychological safety to all employees, and outline expectations for behaviour and communication. Psychological safety is not just a buzzword but also a core value that guides all aspects of organisational culture.

Step 2: Establish Clear Communication Channels

Assess your current communication practices, and identify areas for improvement by gathering open feedback from employees to understand what is really going on in the organisation.

Next, establish structured communication channels that facilitate open dialogue and feedback exchange, whether this is through regular team meetings or town hall sessions, suggestion boxes or platforms for employees to share ideas and concerns. Finally, leaders should actively solicit input and feedback from employees, demonstrate receptiveness to their ideas and suggestions and celebrate contributions regardless of seniority.

Step 3: Promote Psychological Safety in Meetings

Make sure that you establish ground rules for meetings that promote inclusivity, respect and open dialogue. These may include guidelines for active listening, constructive feedback and valuing diverse perspectives. Encourage all participants to contribute their ideas, opinions and perspectives without fear of judgement. Leaders should create an atmosphere where dissenting viewpoints are welcomed and healthy debate is encouraged. You may do this initially through use of assigned roles such as Elephant Spotter, Rabbit Catcher and Giraffe.

Step 4: Provide Training and Support

Regularly provide training programmes and workshops on topics such as communication skills, conflict resolution and emotional intelligence to equip employees with useful tools and strategies and promote a learning culture. Provide support services such as counselling services, employee assistance programmes and mental health resources to support employees cope with stress, anxiety or other challenges that may impact their psychological safety.

Step 5: Measure Psychological Safety

Choose relevant metrics and indicators for measuring psychological safety in the workplace and actually measure them. You might collect these data through employee engagement surveys, turnover rates, exit interviews, absenteeism rates or broader qualitative assessments of team dynamics. Some suggestions for things to measure include:

1. **Employee Net Promoter Score**: This metric gauge assesses employee loyalty and engagement by asking how likely they are to recommend their company as a 'great place to work' on a scale of 0–10. Higher scores correlate with higher psychological safety.
2. **Failure Value:** Ask employees to rate their comfort level admitting or discussing failures. A higher 'failure value' score indicates people feel safe being vulnerable about mistakes.
3. **Candour Metric**: Ask employees to anonymously rate how able they feel to 'bring their full selves' to work and express candid thoughts/opinions. High scores signal psychological safety.
4. **Problem Raising Frequency**: Count how often employees formally log issues, risks or concerns through formal channels. More reporting may suggest a more psychologically safe environment.
5. **Team Psychological Safety Score**: Use quantitative assessments that measure enablers of candour like interpersonal trust and willingness to take risks.
6. **360-Degree Feedback Analysis**: Engage in and review themes from 360-degree feedback about whether an environment feels hierarchical and blame oriented or discourages speaking up.
7. **Exit Interview Themes:** During exit interviews, ask about factors like ability to voice views. Common complaints around stifled voices may indicate psychological unsafety.

The key is establishing a baseline and then continually measuring to identify progress or declines that may warrant interventions. Make sure that you not only analyse but also act on data to inform decision making and prioritise initiatives that support a culture of psychological safety.

Psychological Safety and Culture

Psychological safety is not just a nice-to-have perk, nor is it only a luxury available to certain organisations and even more so it isn't an example of being 'too liberal' or 'feelings focused'. In the words of Amy Edmondson, "Psychological safety is not about being nice; it's about giving candid feedback, openly admitting mistakes, and learning from failure". Psychological safety is an essential ingredient for driving innovation, collaboration and organisational growth, when people feel safe to take interpersonal risks, speak up with candour and freely contribute their unique perspectives. Teams and companies operating with high psychological safety simply perform better. They learn faster from failures, make smarter decisions through healthy debate and keep their finger on the pulse of emerging issues. Rather than wasting energy on obfuscation and self-censorship, people can direct their full talents towards creative problem-solving.

The good news? While cultivating psychological safety takes consistent effort, the strategies for doing so don't have to be complex or costly. As we've seen, simple interventions like modelling vulnerability as a leader, making small gestures to build trust or just asking better questions can start shifting the cultural tone. Ultimately, fostering psychological safety comes down to open communication and human factors like mutual respect, empathy and humility. Whether through incremental habit adjustments or more comprehensive culture change initiatives, working to develop psychological safety demonstrates a significant return on investment. A workplace where people can bring their authentic, candid selves to the table is simply a better one – for employees as well as companies. Perhaps most importantly the level of psychological safety is one of the key factors that will decide whether you approach to cultivating a new sustainable culture it's going to stick.

Cultivating Safety: Activities for Developing Psychological Safety at Work

As well as enacting some of the strategies outlined above for supporting a culture shift towards greater psychological safety, I have described below some activities that can be used to support organisations and teams to reflect on, and develop, psychological safety at work.

Personal Activity	Psychological Safety & Reflection
Objective	The objective of this exercise is to assess workplace psychological safety through direct observation and reflective analysis. By observing team interactions and behaviours in real time, participants can gain insights into the level of psychological safety within the workplace environment.
Duration	15–30 minutes
Materials Needed	Something to write notes on and record observations.

Here's how it works:

Set Up: Select a designated time and location within the workplace to conduct observations. Choose a setting where team members naturally interact, such as team meetings, idea generation sessions, or informal gatherings.

Next define specific criteria or indicators that reflect psychological safety, such as:

• Frequency of speaking up and sharing ideas
• Openness to feedback and constructive criticism
• Inclusivity and respect for diverse perspectives
• Level of trust and collaboration among team members

(Continued)

Observation: Act as an observer during the designated time period, paying close attention to interactions, communication patterns and nonverbal cues among team members. Take notes on observed behaviours and interactions that align with or deviate from the defined criteria for psychological safety.

Reflection & Analysis: After the observation period, take time to reflect on the observed behaviours and interactions. Consider the following questions:

- What patterns or trends did you observe regarding psychological safety?
- Were there instances of individuals speaking up, sharing ideas or offering feedback openly?
- Did team members demonstrate respect for diverse viewpoints and encourage participation from all members?
- How did nonverbal cues, such as body language and tone of voice, impact perceptions of psychological safety?

Identify Strengths & Areas For Improvement: Based on your observations and reflections, identify strengths and areas for improvement related to psychological safety within the workplace. Consider factors such as leadership behaviours, team dynamics and organisational culture that may influence psychological safety.

Action Planning: Develop action plans and strategies to enhance psychological safety based on the identified strengths and areas for improvement. Encourage collaboration and input from team members in co-creating solutions that address underlying issues and promote a culture of trust and openness.

Follow-Up & Evaluation: Follow up the action plans and initiatives developed as a result of the observation and reflection process. Monitor progress, solicit feedback from team members and evaluate the impact of interventions on psychological safety over time. Adjust strategies as needed based on ongoing observations and feedback.

Through direct observation and reflective analysis, participants can gain valuable insights into the level of psychological safety within the workplace environment. By identifying strengths and areas for improvement and developing targeted action plans, organisations can take proactive steps to foster a culture of trust, openness and collaboration that empowers individuals to thrive and succeed.

Team Activity	Sharing Vulnerabilities
Objective	The objective of this exercise is to create a safe space for team members to share personal experiences and vulnerabilities, fostering empathy, trust and psychological safety within the group.
Duration	1–1.5 hours
Materials Needed	Meeting room or designated space conducive to open discussion, chairs arranged in a circle, timer, paper and pens.

Here's how it works:

Introduction (10 minutes): Introduce the purpose of the workshop. Set expectations of open communication, respect, and confidentiality. Outline the agenda and any other ground rules.

(Continued)

Begin by explaining the purpose of the exercise and its relevance to building psychological safety within the team. Emphasise the importance of empathy, active listening and mutual support in creating a supportive and inclusive work environment.

Shared Vulnerability Circle (30 minutes): Have participants sit in a circle and explain that each person will have a designated amount of time (e.g., 5 minutes) to share a challenging personal experience or vulnerability with the group. Use a timer to ensure equal time for each participant.

Active Listening & Reflection (20 minutes): As each participant shares their experience, encourage the rest of the group to practice active listening, offering supportive nods, gestures and empathetic responses. After each sharing, allow a brief moment for reflection before the next participant begins.

Optional Writing Exercise (10 Minutes): After everyone has shared, provide participants with paper and pens and invite them to reflect on their own vulnerabilities or challenges privately. Encourage participants to write down any insights or reflections that emerged during the exercise.

Group Discussion & Reflection (15 minutes): Facilitate a group discussion on the shared experiences and reflections from the exercise. Encourage participants to discuss common themes, insights gained and strategies for supporting each other moving forward.

Action Planning (10 minutes): Conclude the exercise by facilitating a idea generation session on actionable steps that individuals and the team as a whole can take to foster psychological safety in the workplace. Encourage participants to identify specific behaviours, communication strategies and support mechanisms to implement.

Closing (5 minutes): Wrap up the exercise by thanking participants. Remind them of the importance of continuing to support each other and cultivate a culture of psychological safety in the workplace.

Team Activity	Building Bridges
Objective	This classic team-building exercise offers an opportunity to practice open communication, persistence through challenges and boosting creative confidence within a team environment.
Duration	1–1.5 hours
Materials Needed	Meeting room or designated space conducive to open discussion, imer, Sheets of paper (5–7 per small team of 4–5 people), Pencils/pens, Paperclips (5–10 per team), Tape (1 roll per team), Rulers (1 per team) and Tables to build on

Here's how it works:

Introduction (10 minutes): Introduce the purpose of the workshop. Set expectations of open communication, respect and confidentiality. Outline the agenda and any other ground rules.

(*Continued*)

Divide participants into small teams of 4–5 people. Have each team push a few tables together to create their building space. Explain that each team's challenge is to build a freestanding bridge using only the provided materials that can span a distance of at least 12 inches and bear the weight of a heavy book. Emphasize there are no other rules - the bridge design is up to them to conceptualize and construct through collaboration.

Planning & Construction (30 minutes): Start the timer and give tell the team thy have 30 minutes to plan, construct, and test their bridge. Encourage an iterative process of trying ideas and learning from failures. Circulate during the activity and narrate examples of good team behaviours: exploring multiple ideas, dividing responsibilities, giving constructive feedback, celebrating small wins, learning from setbacks, etc.

Testing (10 minutes): Once the time is up, have the teams take turns testing their bridges by placing a heavy book on them. Celebrate all teams who managed to build a freestanding, weight-bearing bridge according to the criteria.

Discussion & Debrief (15 minutes): Lead a discussion around the experience focusing on the team processes and group dynamics. Ask questions like;

• How did the team embrace ambiguity and evolve their approach?
• What helped or hindered productive collaboration and communication?
• How did the team respond to changes or failures? With defeatism or resilience?
• How might this activity relate to challenges teams face in the workplace?
• What did they learn about creating an environment of psychological safety?

Reflection & Action Planning (10 minutes): Encourage the teams to identify how what they have learned can apply to improving psychological safety at work and come up with action plans to begin enacting these strategies.

Team Activity	Giving Permission
Objective	To create a habit of giving teammates 'permission' to take risks and be vulnerable in a supportive environment.
Duration	1–1.5 hours
Materials Needed	Meeting room or designated space conducive to open discussion, pens and half sheets of paper

Here's how it works:

Introduction (10 minutes): Introduce the purpose of the workshop. Set expectations of open communication, respect and confidentiality. Outline the agenda and any other ground rules.

Writing Down Vulnerabilities (10 minutes): Start by having everyone write down 3 risks or acts of vulnerability they tend to hesitate to do in the workplace due to fears or insecurities. These can span from speaking up in meetings to admitting mistakes to trying new skills. Have them write one per index card.

(Continued)

Redistribute Cards & Sharing (30 minutes): Collect all the pieces of paper, shuffle them and redistribute so everyone is holding 2–3 cards with someone else's 'Vulnerabilities' or 'Risks'.

Explain that each person will take turns standing before the group and sharing the risks on the cards they're holding as if they were their own.

When each vulnerability or risk is read aloud, the entire group enthusiastically affirms it by calling out "Permission granted!" while using a gesture like raising their hands or giving a thumbs up. The person reading the card can then elaborate on what makes this a risk, and the group reflects back why it's actually valuable and should be encouraged.

Debrief (10 minutes): Discuss how they found the exercise, explore the impact of receiving that affirmation and vocal support to take risks. How did it feel to have insecurities normalized and validated? Explore ways the group can embed the practice of giving each other 'permission' to be vulnerable and embrace psychological safety as a daily habit.

References

Catmull, E., & Wallace, A. (2014). *Creativity, Inc.: Overcoming the unseen forces that stand in the way of true inspiration.* Random House.

Duhigg, C. (2016) What Google learned from its quest to build the perfect team. *The New York Times Magazine,* 25 February. Available at: nytimes.com/2016/02/28/magazine/what-google-learned-from-its-quest-to-build-the-perfect-team.html (Accessed: 25 March, 2024).

Edmondson, A. (2019). *The fearless organisation: Creating psychological safety in the workplace for learning, innovation, and growth.* Wiley.

Hastings, R., & Meyer, E. (2020). *No rules rules: Netflix and the culture of reinvention.* Penguin.

Vance, A. (2015). *Elon Musk: Tesla, SpaceX, and the Quest for a fantastic future.* Ecco.

10 Cultivating Culture: Designing Workplace Cultures That Support Staff to Create, Innovate and Thrive

Now is where we start to draw all the previous ideas and practices together to begin to understand how we go about the process of consciously constructing a sustainable workplace 'culture'. Drawing on the previous chapters concerning motivation, values, psychological contracts, resilience and psychological safety, this chapter will explore how organisations can actively design their workplace culture so that it benefits the employees and the organisation. Rather than culture simply being 'the way we do things round here', we will explore consciously engaging with its design and understanding that organisation's culture is the defining character and ethos of 'how things work'. It is one of the deciding factors for how much we feel engaged with the organisation and each other, whether people feel attached to the organisational goals and whether people feel motivated to perform.

Culture: What Is It?

Culture refers to the shared patterns of behaviours and interactions, ways of thinking and understanding in a group or 'system' that are learned through a process of socialisation. It describes the beliefs, values, norms, traditions, customs, history of a group of people, shaping identity and influencing how we make sense of the environment around us. Within any given culture, there are usually explicit elements (e.g., language, customs, behaviours) as well as implicit, unconscious elements (e.g., beliefs, attitudes, values). These shared elements are usually transmitted socially, via social 'rules' or 'expectations' from one generation to the next unless significant environmental or attitudinal changes lead to cultural shift (Schein, 2010).

Culture in the context of a workplace is no different. It represents the shared values, beliefs and behaviours that shape how employees interact and work together, their attitudes towards work, communication and their decision making. It may include formal external policies and practices as well as more informal behaviours. Organisational Psychologist Edgar Schein defined culture as "a pattern of shared basic assumptions learned by a group" that describes "...the correct way to perceive, think, and feel...". As such, a healthy, positive culture where basic assumptions of trust, respect and collaboration

DOI: 10.4324/9781003407577-10

are 'correct' is one that supports operational success as well as contributing to the continued wellbeing of employees (Schein, 2010).

If you are curious about what 'culture' might look like in your workplace, ask yourself these questions:

1. *What are the core values that guide decision making and behaviour in our organisation? How are these values demonstrated in practice?*
2. *What formal policies, procedures and structures are in place that shape how work gets done? How might these reinforce or contradict our stated values?*
3. *How does your organisation celebrate successes and milestones?*
4. *Does your organisation have any 'rituals' or 'traditions' (dress-down Friday, sharing lunches, end-of-the-week progress reports), and what do they say about the things your organisation prioritises or values?*
5. *When conflicts or disagreements arise, how are they typically handled? Is there an open culture of dialogue and feedback?*
6. *In what ways does your physical workspace and design reflect and reinforce our culture?*
7. *How diverse are the personal backgrounds, experiences and perspectives represented across our teams? Where might there be cultural blind spots?*
8. *What organisational stories or legends get passed down? What do they reveal about our priorities, assumptions and unwritten rules?*
9. *How would we describe the typical communication style and interaction norms within teams and across the organisation?*
10. *What processes or practices might be so engrained that we don't consciously examine their cultural implications anymore?*

The key is to approach these inquiries with an open, curious mindset, noticing what your organisational culture explicitly prioritises as well as what it might be implicitly reinforced through habits, norms and 'how we do things round here.'. Once we have started to become conscious of our organisation's culture, we can begin to do things to shape that culture for the benefit of the entire organisation.

Aligning Culture and Motivation

One of the first things we can begin to shape is the impact of culture on motivation. Your organisation's culture has a significant impact on employee motivation, and I think organisations have always been aware of this but have attempted to tag on external things to try and resolve any motivation issues, things like extrinsic rewards or incentives. I think it is only recently that organisations have come to realise that culture is a thing they can intentionally shape and nurture to create something that meets both their needs and the needs of employees. Only now do we realise that an unhealthy culture is maybe something we can do something about.

An unhealthy culture is one that undermines motivation. It is one where employees cannot find the energy to engage with their work and where they hold back and don't fully commit to projects; it's one where deadlines are missed and targets are left unachieved. A healthy culture, on the contrary, is one where motivation factors are aligned directly with the organisation's culture, one where employees feel inspired, engaged and empowered to perform at their best. As such, the strategic value of aligning culture and motivations is clear, and from my perspective, so is the process to do it. However, in order to start, you need to understand what really motivates your employees.

As we explored previously, motivation is a multifaceted concept, influenced by both intrinsic and extrinsic factors. Intrinsic motivators stem from within the individual, such as a sense of purpose, autonomy and mastery, while extrinsic motivators represent external influences on behaviour, such as rewards, recognition and career advancement (Pink, 2009). One of the first things to do to begin to align these intrinsic and extrinsic motivations to your organisation's culture is to cultivate a sense of purpose. Help employees understand why you do what you do, why your company exists and what their work is contributing to. I'm reminded of a story I heard once, whilst facilitating a workshop on values-based leadership. The story goes that during a visit by JFK to the NASA Space Centre, he bumped into a cleaner during his tour of the facility. When he introduced himself and asked what the man's role was, the cleaner replied, "I'm helping put a man on the moon, Mr. President". The cleaner felt personally connected to the greater meaning and purpose of his role. By offering people meaningful work and clearly tying roles and tasks back to your broader mission and impact, we can support employees to find motivation even when they are finding the day-to-day tasks challenging (Sinek, 2009). Think about what matters to your employees and what impact they would like to create and connect that to what you do. The example of Patagonia in Chapter 8 is a good example, where leaders directly connected employee actions and work to positive environmental impact for the world.

You can align both intrinsic and extrinsic motivation by ensuring that your company provides opportunities for personal and professional growth and development. Employees are motivated by opportunities for learning and advancement. You can work to create a culture of continuous learning through investing in accessible training programmes and mentorship opportunities that align with the competencies you want to see in your ideal culture. You can understand that the potential for growth motivates retention. Within this you can also empower staff and promote autonomy. A motivational workplace culture is one where people feel they are trusted to make decisions, take ownership of their work and feel they are contributing to broader organisational goals. We observe this in companies like Netflix, where employees are actively encouraged to experiment, take risks and innovate without fear of failure (Hastings & Meyer, 2020). If greater autonomy is a cultural priority for your organisation, think about how you can embed this through practices such as flexible working and decision-making over how works gets done.

If you want particular behaviours to remain in your culture, reward them. This applies to both for both the good and the bad. Since the seminal behavioural experiments of psychologists Burrhus Skinner (1953) and Albert Bandura (1961), in educating pigeons and children, we have known that 'people learn through observation', and that 'the more you reward a behaviour, the more likely it is to continue'. Therefore, if we are aware of the positive behaviours we want to encourage, we can create a culture of appreciation by acknowledging and celebrating individual and team accomplishments. Implement formal recognition programmes that celebrate employees who exemplify the core values and behaviours of your ideal culture; this will motivate them and others to repeat them. This can form part of creating a collaborative and supportive environment. Employees thrive in environments where they feel valued, respected and supported by their peers and leaders. You can build on this by creating opportunities for staff to provide input on policies, processes and initiatives that impact their work. This cultivates a motivating culture of empowerment. You can then go on to encourage greater collaboration through designing processes that not only encourage but also enable cross-functional projects and knowledge sharing. By doing this, you can satisfy employees' intrinsic motivation around learning and growth.

Finally, you can lead by example. You can understand that the behaviours demonstrated by your leaders will be the behaviours most likely to be mirrored by your staff. You can ensure that it is the positive behaviours, the ones you want in your culture, that get modelled by encouraging leaders at all levels to demonstrate the attitudes and behaviours that embody the desired culture. Maybe not eating their lunch at their desk every day or sending email outside of work hours for a start. When employees see values lived out authentically by those in charge, it motivates them to hold those values themselves.

Integrating motivational factors into your organisation's culture requires the alignment of processes, leadership, environment and organisational systems to amplify the intrinsic and extrinsic motivators that are most meaningful to your staff to achieve the culture you want to promote.

Embedding Values in Culture

By now everyone has got organisational values, but very few use them to their true advantage. Organisational values should serve as the compass that guides behaviours, decisions and actions within the organisation. Yet more often than not, they are simply a list of buzzwords handed down from a marketing team with little or no relationship to the actions, beliefs and values of the employees. If you want to devise values that not only weave into the fabric of the culture you have but also cultivate the culture you want, you need to engage in a conscious process of values alignment: define, communicate, embed, empower and reinforce.

Initially when defining a set of meaningful organisational values, don't simply task HR or the marketing team with coming up with ones that sound good. Actively reach out to and engage with a diverse group of staff and stakeholders in the values definition process. Send out surveys, set up focus groups and support staff to engage in workshops to hash out what they feel the company's values are and what they want them to be. This helps to build buy-in from all levels of the organisation. Alongside this, take a look at your current culture and align your value suggestions accordingly. Ask yourself: What are the principles that are genuinely guiding organisational behaviour and decision making? Are they healthy ones that you want to promote, or could they be re-framed as some more beneficial or accessible? A value of 'Making Loads of Money' is perhaps less helpful than a value or 'Growth' or 'Shared Success'. As I mentioned before, keep your final core values to a manageable number, 3–4 (I personally think the magic number is 3), and make sure they are distinct, memorable and aspirational.

Once you have come up with your core values, think about how you plan to communicate them to the organisation. I remember when working in the National Health Service, I left work on the Friday only to return on the Monday to find A4 posters up in the toilets announcing what our new core values were. This was the first I had heard of them. Consider designing engaging, creative internal communication campaigns to introduce and regularly reinforce the meaning behind each value. Display your shiny new value statements throughout the workspace and on email signatures or business cards to keep them in people's minds. Finally, if you plan to write your values out as statements or guiding principles, be conscious of the language that you use. It is better to frame your values as verbs rather than simply nouns. For example, 'Empowering People' is better than just 'Empowering'. Similarly make your statements aspirational rather than simple statements of what you are doing. For example, 'We Inspire Staff' is less helpful than 'To Inspire Staff'. The reason being that if you offer them as statements of fact, the minute I don't feel inspired, the statement loses its meaning and I lose trust. The idea being that these are things we are aiming for, things we hope to achieve rather than things we have already done.

This is often where organisations stop. They come up with some values and then push them out and miss the next crucial step of embedding them into practice. If you want them to stick and create the sustainable culture you are after, you need to identify opportunities to map them onto existing organisational processes, policies and systems. You can embed value language into your performance management processes, recruitment procedures and onboarding and training materials. You can model the values in practice by having leaders repeatedly reference and illustrate how decisions and initiatives uphold the values, and you can reward value-aligned behaviours in staff formally and informally.

Next empower your staff to understand, embody and take ownership of the organisation's values. You can provide staff with training to develop their awareness and understanding about what the values mean and how they connect to their own values, showing how they play out in their daily actions. You can encourage them to take ownership of the organisation's values by creating formal and informal platforms for them to challenges areas of values misalignment and suggest changes. You can also empower staff to explore innovative solutions and take calculated risks in service of the values, using them as framework for decision making.

It doesn't end there. You can't just push out the values; encourage people to live them, and then sit back and watch your culture magically change. You need to continue to measure and reinforce them. You can do this by including value-aligned metrics in performance evaluations and incentive criteria. You can include measures in your staff surveys to look at the alignment between stated values and staff perceptions of those values. Collecting these data as well as open staff feedback allows you to adjust your practice to better embed these values into your culture over time. Before you say, "This sounds like a tall order" or "It can't be that simple", your company already had values, way before you wrote them out on the piece of paper, but they just weren't formalised. These values were formed unconsciously, and they stayed pretty rigid; otherwise, you probably wouldn't be reading this book. The process I outline above is a conscious one, and if you are consistent, it really is that simple. The key to making organisational values truly meaningful is consistent role-modelling, open dialogue and an unwavering commitment to embodying them at all levels. This takes ongoing work but builds a motivating, principled culture that lasts (Cameron & Quin, 2011).

Designing Psychological Contracts to Influence Culture

From Chapter 4, you should understand that psychological contracts are the secret glue that binds employees to the organisation, influencing their loyalty, engagement and discretionary effort (Rousseau, 1989). They refer to the unwritten set of expectations and beliefs that employees and employers hold about what they owe each other, beyond the formal job description or employment contract. These contracts are extremely powerful, often more so than the formal contract. They shape the overall relationship between an individual and their organisation. When these psychological contracts are well designed and upheld, they have a positive impact on organisational culture, commitment and performance. These contracts are usually either transactional or relational, sometimes a collection of the two. What we want is less of the short-term, quid pro quo transactional contracts and more of the deep socio-emotional relational contracts that improve mutual investment, shared commitment and greater organisational identification over time.

In order to consciously design these contracts for the benefit of your culture, you need to begin by openly exploring the ones you already have. If you are curious about which contracts dominate at your workplace or even between yourself and your workplace, I invite you to reflect on the following questions.

Transactional Contracts:

1. *Is the relationship between the employee and the organisation primarily economic/ monetary in nature? Is the focus primarily on exchanging work for financial compensation?*
2. *Are work expectations defined by a rigid job description with little room for flexibility or evolution of duties over time?*
3. *Is there an underlying assumption that the employee will remain with the organisation only for a limited, defined period of time?*
4. *Do employees feel like they are viewed as disposable resources or expendable once their current prescribed role is complete?*
5. *Is there a lack of trust, commitment or attachment between the employee and the broader organisational mission?*

Relational Contracts:

1. *Is there a sense of open-ended investment and commitment between the employee and the organisation that extends beyond a simple economic exchange?*
2. *Are there opportunities for employees to expand their role, grow their skills and advance their career over time?*
3. *Does the organisation provide professional development, mentoring or supportive services beyond just salary/benefits?*
4. *Do employees feel a sense of meaning, purpose and emotional connection to the organisation's mission and values?*
5. *Is there a foundation of trust, open communication and good faith that the employee and organisation will be responsive to each other's changing needs and circumstances?*
6. *Are there formal and informal mechanisms for employees to voice concerns, engage in organisational decision making and help shape policies/practices?*

By openly discussing the current state and desired nature of the psychological contract, you can better understand whether changes are needed to cultivate a more resilient, engaged and committed workforce. You can do this actively through open discussions with employees about their understanding of the mutual expectations of their role and matching those with the reality of what you expect to give and receive in line with your company's values.

To ensure that these conscious contracts are upheld, it is important that you go on to actively build trust and promote transparency, ensuring consistency between your stated values or promises and your actual practices. You can encourage leaders to communicate openly and honestly and demonstrate you care for employee wellbeing. If the psychological contract includes expectations of development and growth, make sure this happens or the perceived violation of the contract will lead to disengagement. Make sure that you provide learning opportunities, aligning them with employee interests, and provide a clear and transparent process for career development. Take efforts to empower your staff by giving them a voice through forums and follow up on any concerns or suggestions where possible.

The majority of healthy psychological contracts will include expectations or respect and fairness. You can meet these expectations through fair and equitable policies and systems with regard to rewards and recognition. Celebrate diversity or thought, identity and experience, and cultivate a culture of flexibility, meeting employees where they are before supporting them to get where we need them to be. Remember that psychological contracts already exist in your organisation and always have, and they have been silently shaping engagement and the essence of workplace culture for years. By proactively shaping and delivering on a healthy employee–organisation relationship, you build an organisational culture of trust, motivation and reciprocal investment. This serves as the 'psychological glue' for commitment.

Cultivate a Resilience- and WellBeing-Oriented Culture

Another set of cultural factors we can influence are resilience and wellbeing. These are essential for a healthy and thriving workplace as they enable employees to navigate challenges, adapt to change and manage stress. I will explore the business case for 'Wellbeing As A KPI' in Chapter 12, but with regards to culture cultivation, the aspects we can directly influence are openness, work–life balance, self-care, community support and continuous assessment.

As we discussed earlier, resilience refers to our ability to adapt to and overcome change. In order to do that, we need a culture of self-awareness and open communication. A wellbeing-oriented culture is one where people can talk openly about their mental health and wellbeing without fear of stigma or reprisal. When was the last time you asked someone at work "How is your mental health?" If you have a culture that has space for this already, then great; if not, then think about how you can encourage open communication. Whether it's through training in mental health awareness, stress management or training for leaders around supportive conversations or through having employees share personal stories of resilience. What you want is a culture where the fear of talking about mental health is removed so that people are able to reach out when they need support or when they are struggling. Remember this is easier said than done and will likely need an organisational wide approach to shifting communication. If you are curious how good your staff are at saying when they are stressed, take a look at your HR data around the number of staff absences due to 'upset tummies'. Then compare those data to the general population. No doubt you will discover that there is a disproportionate number of upset tummies at your work. Why? Because people are using other ways to communicate distress. It seems it is easier to talk about diarrhoea than stress.

As well as inviting people to communicate their distress, we can work to build up their resilience and ability to tolerate stress. The simplest way to do this is a culture that promotes work–life balance and encourages active self-care. Promoting true work–life balance in an organisation is often easier said than done, especially when it is at odds with your organisation's existing culture.

Imagine some leaders at a company who have heard about the importance of work–life balance at an inspirational conference they attended last week. They decide to send out a suitably inspiring email saying, "From now on, we're promoting better work-life balance! No more working nights and weekends – we value your personal time!". The email is met by a mix of relief and scepticism from the employees. Susan, in accounting, thinks: "That's nice in theory, but they still expect me to have these financial reports done first thing Monday". Paul, a manager himself, worries: "That's all well and good, but it doesn't change the targets for this quarter.". A month goes by, and John, a dedicated employee, decides to be bold and ignore his work email over the weekend to spend time with his family. He returns Monday, refreshed, and is asked if he "has a minute" by his team leader and promptly receives a talking-to about being "unresponsive" over the weekend and "taking his eye off the ball.". John goes back to business as usual, feeling disappointed and let down.

Simply expressing support for work–life balance without concrete changes to ingrained practices and leadership behaviours is unlikely to stick. Some tangible steps you can take to enact real change include:

1. **Implement Work–Life Boundaries:** Establish no-meeting periods like nights/weekends. Have leaders role-model by not sending late-night emails.
2. **Provide Flexibility**: Offer flexible scheduling, remote options and generous paid time off so that employees don't feel forced to work extreme hours.
3. **Discuss Priorities**: Have team conversations about realistic prioritisation – what can be reasonably accomplished in a healthy workweek?
4. **Focus on Sustainability**: In periods of intense crunch, ensure it's followed by true downtime to prevent chronic overwork.
5. **Train Managers**: Help them understand the costs of burnout and how to notice signs, delegate more and prioritise recovery. Remember, happy bees work harder.

The key to making sure that employees' work and life are genuinely in balance requires more than just words; it requires redefining processes, updating leadership behaviours and re-designing the way 'work' gets done. And it matters immensely because chronic workplace stress contributes to absenteeism, turnover, health issues and £ millions in costs per year. A culture of balance isn't just good for people; it's good for business too.

The same approach is true of self-care, seemingly another corporate buzzword of our time. Picture this all-too-familiar story. Imagine a mid-sized tech company with a high-intensity culture. One of the senior leadership team has been reading some articles about the importance of employee self-care for overall wellbeing and productivity. Feeling inspired, they bring in some self-care experts to provide training to the staff on stress management techniques like meditation, yoga and mindfulness. During the training session, employees nod along and think "Yes, this all sounds wonderful. I could really use some tools to manage stress and prevent burnout". However, participants are dubious given the company's culture.

Sure enough, a few weeks later, Michael decides to actually take advantage of the suggested self-care practices. He starts taking mindfulness breaks every couple of hours and even practises some deep breathing exercises before opening his email in the morning. His manager Jenny quickly pulls him aside with a puzzled look: "Michael, I noticed you've been stepping away from your desk a lot lately. Is everything okay? We really need your focus to support the team on the current project.". Similarly, across the office, a new manager, Francis, tries to convince the team to implement a Friday self-care hour when they can relax, decompress and recharge. But their team members just laugh and say, "As if! I can barely find time to use the bathroom, let alone a self-care hour!"

While well-intentioned, the company's words about supporting self-care rang hollow because the cultural realities made it very difficult for employees to actually prioritise self-care without feeling guilty or facing negative repercussions. To cultivate a self-care-oriented culture, organisations need to go beyond just providing training and take concrete actions, such as:

1. **Destigmatising self-care** by openly discussing and role-modelling it at a leadership level. Make wellbeing support services accessible such as counselling or the Employee Assistance Program.
2. **Auditing workplace practices** and processes that cause excessive stress/burnout and addressing them.
3. **Building in formalised break times** and opportunities for employees to practise taught self-care activities.
4. **Rewarding and celebrating employees** who effectively practise self-care while sustaining high performance.
5. **Continuously communicating** about and improving self-care supportive policies/benefits over time. Show your commitment to the culture.

The difference is making self-care an integral, expected and rewarded part of how work gets done rather than a nice bonus or something expendable. Actively engaging in self-care allows employees to sustainably work at their highest levels without burnout. It's an investment that pays dividends through improved morale, retention and overall organisational health.

By now you should appreciate that these strategies don't work in isolation. Encouraging self-care and work–life balance in individuals needs to be maintained within a supportive and resilience culture. Building a strong supportive community at work is critical for creating an environment where employees feel connected and engaged. Without proactive efforts to build supportive communities, even well-intentioned workplaces can struggle to retain top talent. You can embed a culture of connection in a number of ways, many of which you might already be engaged in:

1. **Peer Support Programs**: Formalise peer support systems where trained employees provide a listening ear and assist colleagues navigating challenges. This destigmatises struggles.
2. **Mental Health Champions**: Identify and equip volunteer employees to serve as mental health ambassadors, promoting resources and creating safe spaces to have courageous conversations.
3. **Employee Resource Groups**: Encourage employees to create communities based on common interests, identities or experiences. These foster connection beyond just work roles.
4. **Social Events**: Intentionally organise events and affinity groups centred on fun social connection and bonding rather than strictly work activities.
5. **Integration Activities**: Design creative exercises as part of team meetings to learn about each other's backgrounds, strengths and interests beyond just their job duties.

Design proactive, creative and structured opportunities for employees to form meaningful relationships and find that sense of community at work. This, in turn, increases engagement, openness, trust and commitment to the overall workplace culture.

Finally, once you have begun to orientate towards a wellbeing-focused culture, you need to continuously assess and improve. You can build in your strategies and let them go, but you need to make sure they are working as intended from time to time. You can do this through collecting employee data on stressors, pain points and needs through employee wellbeing surveys. If you are trying to design a culture that improves wellbeing, then measure wellbeing, or resilience or satisfaction or whatever it is. Notice where it has worked and where it hasn't, identify what you could do differently, seek feedback and try something else. It may take a few tries before you get it right.

Remember that this process takes time, and the most ingrained cultures are often the hardest to shift. The process of designing and constructing workplace culture is dynamic and multifaceted and should never been approached as an ad hoc or reactive process. It is one that requires intentional effort and strategic alignment to do well, and it is one that only works by embedding motivation, values, psychological contracts and wellbeing orientation into the very fabric of how your organisation operates. The holistic approach to cultural integration is what allows a positive culture to become truly sustainable.

How to Shift a Culture

I won't lie to you; consciously shifting an organisation's culture towards one that is healthier and more sustainable is a significant undertaking but very much worthwhile. In general, there are two approaches organisations can take, either an incremental step-by-step process or a more comprehensive, large-scale cultural transformation.

The incremental approach involves implementing small culture changes over an extended period of time. This may start with some pilot programmes, updating certain policies, providing training to specific teams and making moderate adjustments to processes. The benefits are that it feels less disruptive, allows for small wins to help bolster the process and provides time to gather feedback and adjust the programme as needed. The downsides are that it can take a very long time to see significant cultural change, momentum can get lost and the incremental tactics may be easily disruptive by unexpected organisation change from other sources.

Alternatively, pursuing an all-in, large-scale cultural overhaul from the outset is often more appealing to the bold and decisive. The comprehensive restructure involves rapidly redesigning major systems like leadership philosophies, workplace practices, incentive structures, physical environments and employee resources to reflect the cultural shift that you are looking to achieve. This approach allows more immediate, broadscale progress; it is often very costly, requires impeccable planning and is more likely to be met with higher levels of resistance.

In my opinion, it is a variation on the first option that works best. This is based on the awareness that organisational cultures shift more readily from the edges. An edge-driven approach is often cheaper, easier to mobilise and can provide 'case studies' to encourage deeper cultural shift across the entire organisation. Firstly, make sure to map your edge-driven approach to the qualities you are seeking to develop in your culture. For example, if you are looking to inspire innovation, then give specific groups within the organisation freedom to experiment with new ideas and practices and give them cover to take risks. Find examples on the periphery of your organisation where individuals or groups are showcasing the cultural qualities you are looking to develop and celebrate and amplify their success across the organisation: "If they can do it in accounts, maybe we can do it here?" Find examples of groups or teams within your organisation who are already demonstrating the broader cultural qualities you are after, provide opportunities for the leaders of that change to champion those stories and provide resources to implement those strategies across the organisation. Finally engage in 'No Fail Experiments'; try out something new in a peripheral group, a new way of working perhaps, and approach the experiment with the mindset that the only outcome is learning. In this instance, it can never be a failure as whichever way it goes, you will learn something.

Culture change is often challenging when we attempt to drive it from the top; however, if we can find ways to give it fertile ground to take root organically from the peripheries, this can prove a far more powerful catalyst for sustainable change. Regardless of the approach you decide to take, any substantial organisational culture change is likely to be met with setbacks: whether there is a lack of resources to maintain momentum or active opposition from sceptical employees or old-school diehards wedded to the status quo. Teams may be concerned about the potential impact the transition may have on productivity or that the cultural values you are seeking to promote are the 'wrong' ones.

To overcome these obstacles, it's crucial to approach culture shift from a coherent, committed and collaborative perspective, engaging employees from all levels of the organisation. Find the change-makers, the culture champions who are already engaging in the behaviours and practices you want to cultivate. Make them visible cultural ambassadors whom people can relate to and who can help champion your cause. Most importantly, prioritise feedback, open communication and listening to how your employees feel, and be prepared to adapt as you go. Yes, the process of consciously shifting a culture to a healthier, more sustainable one is challenging; however, if we approach it intentionally, it is something an organisation of any size can achieve. By taking a thoughtful, sustained and inclusive approach, even the most deeply entrenched cultures can successfully transform. In the end, a healthier, more sustainable culture provides everything you need to improve wellbeing, retention, resilience, connection, creativity, motivation and performance. You already have a workplace culture, but do you have a workplace culture that works?

Evaluating Culture: Organisational Culture Assessment

Exploring organisational culture is best achieved through asking those directly affected by it, the staff.

Team Activity	*Organisational Culture Assessment*
Objective	To gain insights into the organisation's current culture, identify areas for improvement, and collaboratively develop actionable strategies to cultivate a more positive and supportive work environment.
Duration	1–1.5 hours
Materials Needed	Meeting room or designated space conducive to open discussion, flip chart paper or whiteboard, markers, sticky notes and pens.

Here's how it works:

Introduction (10 minutes): Introduce the purpose of the workshop. Set expectations of open communication, respect and confidentiality. Outline the agenda and any other ground rules.

Briefly introduce the purpose of the exercise: to assess the current organisational culture and identify strengths and areas for improvement. Emphasise the importance of honest and constructive participation from all team members.

(Continued)

Idea Generation (10 minutes): Divide the group into smaller teams of 3–5 members.

Each team generate ideas and discusses key aspects of the organisational culture they have observed or experienced. Encourage participants to consider elements such as values, communication, decision-making processes, leadership style and teamwork. Ask them to consider the question "How do we do things around here?".

Identify Strengths & Weaknesses (15 minutes): Each team writes down their observations on sticky notes, separating them into strengths and weaknesses categories. Stick the notes on the flip chart or whiteboard, clustering similar points together.

Group Discussion (20 Minutes): Facilitate a group discussion on the strengths and weaknesses identified. Encourage participants to share examples or anecdotes that support their observations. Discuss the potential impact of these strengths and weaknesses on employee morale, productivity and overall organisational performance.

Prioritisation (10 minutes): As a group, prioritise the top three strengths and top three weaknesses identified. Discuss why these areas are considered most important and how they contribute to or hinder the organisation's success.

Action Planning (15 minutes): Identify potential actions or initiatives to leverage strengths and address weaknesses. Assign responsibility for each action item, and establish timelines for implementation. Emphasize the importance of accountability and follow-up to ensure progress.

De-Brief & Reflection (5 minutes): Summarize the key takeaways from the exercise.

Invite participants to reflect on what they learned about the organisational culture and how they can contribute to positive change moving forward.

References

Bandura, A., Ross, D., & Ross, S. A. (1961). Transmission of aggression through imitation of aggressive models. *The Journal of Abnormal and Social Psychology*, 63(3), 575–582.

Cameron, K. S., & Quinn, R. E. (2011). *Diagnosing and changing organisational culture: Based on the competing values framework*. John Wiley & Sons.

Hastings, R., & Meyer, E. (2020). *No rules rules: Netflix and the culture of reinvention*. Penguin.

Pink, D. H. (2009). *Drive: The surprising truth about what motivates us*. Riverhead Books.

Rousseau, D. M. (1989). Psychological and implied contracts in organisations. *Employee Responsibilities and Rights Journal*, 2(2), 121–139.

Schein, E. H. (2010). *Organizational culture and leadership* (4th ed.). Jossey-Bass.

Sinek, S. (2009). *Start with why: How great leaders inspire everyone to take action*. Penguin.

Skinner, B. F. (1953). *Science and Human Behavior*. Macmillan.

11 Systemic Leadership: Diagnosing and Making Change at an Organisational Level

One of the clearest examples of the 'Us' in the transpersonal perspective is found in the orientation towards systemic leadership. In this chapter we will draw together the knowledge and ideas from previous chapters under the umbrella of systemic leadership. I will offer a definition of the term, its core constructs, how we use it every day and in workplace dynamics, and define its purpose in taking accountability for the work of the organisation as a whole/as a system. Many of the core skills of a systemic leader have been covered already; self-awareness, having an open mindset, attention to diversity and equality, the value of relationships and trust, effective communication and the co-creation of structures and strategies. Here we will explore how we can hold an understanding of the system as a whole in mind to guide decision making, and the conscious cultivation and enactment of a healthy workplace culture.

Systemic Leadership: What Is It?

I should probably start with a definition of the term 'Systemics'. It may be something you are familiar with. It was born primarily from fields such as computational neuroscience and is an approach that focuses on understanding how different parts of a system/group/organism/machine interact with each other and influence the behaviour of the system as a whole. It recognises that systems are interconnected and that changes in one part can have knock-on effects across the rest of the system.

As an example, I'm not sure where you are reading this, but, if possible, go to a window and look out to see the nearest green space or garden. In a space like this, the plants, soil, water, sunlight and even insects and microorganisms are all interconnected parts of the system. If one part is out of balance, such as too much or too little water, it can disrupt the entire system and affect the growth and health of the plants. In the context of an organisation, systemics recognises that different departments, teams, processes and individuals are interconnected parts of a larger system – 'the company'. A change in one area, such as a new policy, practice or technology, can have ripple effects

DOI: 10.4324/9781003407577-11

on other areas, such as employee morale, customer satisfaction or financial performance (Meadows, 2008).

The key principle of systemics is that systems cannot be fully understood by looking at their individual parts in isolation. If you took a bicycle to pieces and examined each of the parts individually, not knowing what a bicycle was, would you be able to discern their function and particularly their function as the 'whole' bicycle? Instead, it is necessary to examine the relationships, interactions and interdependencies between the various parts to understand the behaviour and dynamics of the system. We notice that the pedals turn the cog, which turns the chain, which moves the wheel and so on.

In systemic leadership, leaders work to adopt a systemic approach, seeking a deeper understanding of the complex relationships, interconnectedness and interdependence within their organisational system, enabling them to make more informed decisions, anticipate unintended consequences, develop more effective strategies for achieving goals and navigate the VUCA (Volatile, Uncertain, Complex and Ambiguous) age (Wheatley, 2006). The core principles of a systemic leadership framework are engaging in systems thinking, encouraging collaboration, promoting adaptability and resilience, and having a shared vision and purpose. Some of these qualities and characteristics you will be familiar with from previous chapters; some perhaps less so.

The first crucial skill to grasp is 'systems thinking'. This describes the ability to successfully view an organisation as a dynamic and interconnected web rather than as a series of isolated parts. Like in the phrase "The whole is more than the sum of its parts". Systemic thinkers understand that complex problems are not solved by focusing on individual components and see systemic solutions, acknowledging the interdependencies in the system, to make decisions and address underlying patterns influencing the problem. When we understand how different parts of a system interconnect and influence one another, we are able to see the 'bigger picture', something people often talk about but rarely get a true glimpse of (Gharajedaghi, 2011).

Take your morning routine for example. What is your process? Get up, do some exercise, have a shower, eat breakfast, get ready? When everything goes as per plan and the system interacts as it should, I imagine you feel better, more energised and have a good start to the day. If you were to skip breakfast, get a bad night's sleep, be late and miss the gym, you might be left feeling low, sluggish, hurried, and it would affect your mood and performance for the start of the day and maybe beyond. A systems thinker would notice how all these things are connected in their morning routine and contribute to their overall mood. For a more complex example, let's consider you are booking a family holiday. Rather than simply focusing on the desired outcome of achieving 'Holiday' and simply picking a destination, booking a hotel and flights, a systems thinking approach would consider the various interconnected elements that could impact the overall experience. These might include family dynamics, where you consider the needs and preferences of all the family members as well as potential sources of stress of reach. You might then adjust your choices to better accommodate those needs. A systems thinker may consider transportation. How will you be getting to and from the airport, for example? What about airport parking, rental cars or public transportation options? You might think about the accommodation: what is its location in relation to attractions, restaurants and other amenities that might impact your ability to engage in specific activities. Finally, what about budget? Instead of just looking at the cost of flights and hotel, a systems thinking perspective might take a holistic view of your overall vacation budget, including factors like dining, activities, souvenirs and unexpected expenses. Consideration of all of which may lead to a more positive outcome for everyone involved, avoid potential issue and increase the likelihood that you might want to do it again. Systems thinking in everyday life involves stepping back and looking at the bigger picture rather than focusing on isolated parts or tasks. It helps you anticipate ripple effects, identify potential conflicts and make more holistic decisions that consider the various interconnected elements of a situation.

Bringing it back to the workplace, imagine you're a leader who has been tasked with improving employee productivity and engagement. Usually in a situation like this, people focus on individual performance or implement isolated initiatives such as a 'time management' or 'productivity' training course (I know, I've been asked to deliver my fair share). A systems thinking approach would involve considering the various interconnected elements that contribute to a productive and engaged workforce. Elements include:

1. **Organisational Culture:** How do the company's values, norms and underlying beliefs shape employees' behaviours and attitudes? Does this organisation promote positive communication, collaboration and decision-making processes? Is the problem 'Us', not 'Them'?
2. **Leadership and Management**: Is your organisation's leadership effective in providing support and guidance and encouraging a positive work environment? Are there negative aspects of the leadership culture that are influencing engagement and motivation?
3. **Work Processes and Systems:** Rather than just looking at individual tasks, you would evaluate the overall work processes, tools and systems that employees interact with daily. Maybe you need to streamline workflows, eliminate inefficiencies and provide the necessary resources to support productivity.
4. **Employee Development and Growth**: Providing opportunities for learning and development can contribute to employee engagement and motivation, which, in turn, will contribute to productivity. A systems perspective might consider assessing your current training and development programmes, as well as identifying pathways for growth and progression within the company.
5. **Work–Life Balance and Support Systems**: Recognising that employees have meaningful lives outside of work and these may be impacting their productivity and engagement. A systems thinker would consider

the support systems in place to promote work-life balance and wellbeing and whether improvement is needed.

6. **Feedback Loops and Continuous Improvement**: A culture of continuous learning and improvement is one that inspires greater motivation and accountability. A systems approach would involve ensuring that mechanisms are in place for gathering employee input, identifying areas for improvement and implementing changes based on that feedback.

By considering all these interconnected elements, you can develop a more comprehensive and effective strategy for improving employee productivity and engagement – one that takes into account all the elements of the system that may be influencing it. It allows you to identify potential bottlenecks, leverage connection between different aspects of the workplace environment and make more informed decisions that address the root causes of the productivity or engagement challenges.

The remaining qualities we have covered in previous chapters. Systemic leaders recognise that no single individual or department can solve complex organisational problems alone. So instead, they foster collaboration and engage stakeholders from various levels and functions within the organisation and externally in seeking solutions. Systemic leaders also understand the importance of adaptability and resilience. They embrace change, learn from experience and adjust their strategies and approaches as needed to ensure the organisation's long-term viability. This ability to consider the longer-term goals, aspirations and needs of the system is echoed in their ability to articulate a clear and compelling vision for the organisation, a vision that serves to aligning a system around a common purpose and a shared sense of meaning.

Systemic Problem-Solving

I've mentioned the role of systems thinking in finding solutions to complex problems. The reality is that the majority of problems in any large workplace are likely complex in nature and influenced by a range of elements within the system. Therefore, finding the 'root cause' of an issue can be a challenge. This is another area where a systems thinking perspective can be useful. A perfect example of this in practice can be found in the famous 'Five Why's' technique of Sakichi Toyoda, the founder of Toyota Industries. Here, when faced with a 'problem', you repeatedly ask the question "Why?" in an effort to go beyond surface-level explanations to uncover the underlying cause. It serves as a useful tool for understanding the interconnected nature of problems within complex organisational systems (Liker, 2004).

There is a fantastic example of real-world example of this technique at work, which has become almost apocryphal within the 'five-whys' literature. This is the story of the Jefferson Memorial and the findings of an unpublished research study by Professor Donald Messersmith (1993). The story goes that managers at the national park were concerned that the Jefferson Memorial in Washington DC was deteriorating more rapidly than other statues. They began to ask why:

Why 1: Why is the monument deteriorating faster than others?
Because it is being frequently cleaned with harsh chemicals.
Why 2: Why is it being frequently cleaned with harsh chemicals?
Because of the large amount of bird poo that regularly covers the monument.

They could have stopped there and offered a solution of less cleaning perhaps. However, this would have meant a statue regularly coated in bird poo. They could have stopped at the birds and thought about ways to manage them; introduce falcons, fix lots of spikes to the statue, employ a scarecrow, but this would have incurred a cost and introduced new issues. They could have considered moving the statue to another part of the park, but this would have come at a cost and again may not have solved the problem. Instead, the park managers kept asking why and with the help of Professor Messersmith sought the root cause.

Why 3: Why is there regularly a large amount of bird poo on the monument?
Because the birds are attracted to the high number of spiders that live on or around the monument to eat as food.
Why 4: Why are there so many spiders in and around the monument?
Because of the swarms of insects around the monument, particularly in the early evening, that serve as food for the spiders.
Why 5: Why are there so many insects swarming around the monument, particularly in the early evening?
Because the surrounding lighting that comes on in the early evening attracts insects to the monument.

The solution therefore becomes 'changing the lighting on the monument in the early evening to reduce the number of insects that swarm around it'. A simple cost-effective solution that solves the problem on a systemic level. Less insects – less spiders – less Birds – less Poo – less Cleaning = less damage to the monument.

Thinking about this process in a workplace context, let's consider the steps to go about using the five-whys technique to explore an organisational challenge:

Step 1 Identify the problem or issue. Firstly, you need to define the problem that needs to be addressed. For example, increased absence, low employee engagement, poor customer satisfaction or declining productivity.
Step 2 Ask "Why?" repeatedly. Once you have identified the problem, start asking "Why?" to uncover the underlying causes. Each time a potential cause is identified, the question "Why?" is asked again to dig deeper into the root cause. In the example of low engagement: 1) "Why aren't

people engaged?" – "Because they feel undervalued and unmotivated". 2) "Why …?" – "Because there is a lack of recognition and growth opportunities". 3) "Why …?" – "Because the career development programmes are unclear". 4) "Why …" – "Because there is a disconnect between HR policies and department needs". 5) "Why …" – "Because of siloed decision making and a lack of collaboration".

Step 3 Uncover the systemic issues: By repeatedly asking "Why?" and digging deeper, we can uncover systemic issues that may be contributing to the problem. In the example above, the root cause may be siloed information and disconnect between departments.

Step 4 Identify interconnection. As the root causes are uncovered, systemic leaders can identify interconnections and relationships between different parts of the system that may be contributing to or making the problem worse. This holistic understanding is crucial for developing effective and sustainable solutions.

Step 5 Develop systemic solutions. Once you have a deeper understanding of the root causes and interconnections, systemic leaders can develop solutions that address not just the surface-level symptoms but also the underlying systemic issues.

This technique aligns with the principles of systemic leadership by encouraging leaders to look beyond surface-level 'symptoms' and to be curious about the underlying interconnected factors that might be contributing to the problem. By uncovering root causes and systemic issues, leaders can develop more comprehensive and effective solutions that address the underlying issues in the system as a whole.

As such, the value of systemic leadership lies in its ability to address complex challenges and create sustainable solutions. By recognising the interconnectedness of complexity of various elements within the system, systemic leaders can make more informed decisions, anticipate potential consequences and navigate through uncertainty and ambiguity, to help resolve issues at the root rather than sticking plasters over the top.

Systemic Leadership: What Does It Look Like?

Back to your imaginations again. Picture yourself in the bustling offices of BeeHive.Inc. Employees are buzzing about working busily on their tasks independently. However, despite their individual efforts, they often find themselves facing bottlenecks and inefficiencies in their workflows. One day, Systemic Sandra, the company's CEO, decides to shake things up. She gathers the team together to explore the concept of systemic leadership. She explains that instead of focusing solely on their individual tasks, they need to think about how their work fits into the broader organisational system. Using the analogy of a well-oiled machine, she encourages them to consider how each component of the organisation interacts with the others to achieve common goals. Inspired by her words, the team begins to collaborate more closely, sharing ideas and resources to streamline processes and overcome obstacles. They implement new communication channels, break down silos between departments and adopt a more holistic approach to problem-solving.

Before long, the office transforms into a hive of productivity and creativity. The employees work together seamlessly, leveraging their collective strengths to drive innovation and success. With Sandra's guidance and the power of systemic leadership, they overcome challenges with ease and create a workplace culture where everyone feels valued, supported and empowered to make a difference. Here the power of systemic leadership shines through the chaos to transform the workplace into a harmonious hive – if only it were that simple.

Yes, maybe that is an unrealistic example. As we know from before, systems can be hard to shift, and ingrained cultures and behaviours ever more so. Successfully implementing systemic leadership approaches requires your leaders to possess or develop a systems thinking mindset, characterised by open-mindedness, considering longer-term implications of actions, seeking diverse perspectives, embracing complexity, noticing patterns and a willingness to adapt. As a leader or in any other role, you can assess your ability to engage in systems thinking by asking yourself these questions:

1. *When faced with a problem or challenge, do I take a step back to understand the broader context and wonder how different elements might be influencing each other?*
2. *Am I open to challenging my own assumptions and considering alternative perspectives?*
3. *Can I identify recurring patterns or trends that may indicate underlying systemic issues?*
4. *Do I anticipate how different parts of the system might react or adapt to the proposed changes?*
5. *Am I comfortable with ambiguity and complexity, or do I tend to oversimplify situations?*
6. *Can I hold multiple perspectives and variables in mind simultaneously?*
7. *Do I actively involve stakeholders from different parts of the system in problem-solving and decision-making processes?*
8. *Am I willing to question and challenge existing mental models and assumptions?*
9. *Do I actively seek feedback and incorporate lessons learned to adapt my approach over time?*
10. *When making decisions, do I consider the long-term implications and sustainability of the proposed solutions?*

If you answered each of these questions with a resounding 'Yes!', then you are already stepping into the systems thinker mindset.

In order to view the system as a whole, you have to be open to seeing it that way, to understanding that solutions to complex problems often require innovative thinking and cognitive flexibility. It's this openness to new ideas that will support leaders to embrace a systems thinking model and effectively navigate uncertainty with creative problem-solving. An open-minded leader is one who is open to new ideas, to experimentation, to understanding that doing the same thing leads to the same thing and only change leads to change. They also need to have a vision for where that change is leading. Rather than focusing on short-term wins, they need to be able to hold in mind longer-term objectives that align with the organisation's vision and values. What is it they are hoping to achieve in the long run, and what are the steps they need to take to get there? They need to have the requisite imaginative capacity to see beyond what is good for the company, or more likely themselves or their team, now and what it will need in the future.

Somebody once told me about a philosophy called the '7th Generation Principle', attributed to the Iroquois, that describes how decisions we make today should result in a sustainable world seven generations into the future. This philosophy seems extremely prescient given our current global environmental struggles and, I think, applies just as well to the successful cultivation of a sustainable culture in an organisational system. A systemic leader considers not only the needs and motivations of the system now but also how their actions will meet the needs of the system in the future. To support this open-minded, future-focused perspective, a systemic leader needs to actively seek out diverse perspectives and novel ideas. By fostering an environment that champions diversity and inclusivity, we can create systems where all individuals feel respected and valued and able to share their unique viewpoints. This serves to expose leaders to new ideas while also enhancing creativity and collaboration. In turn, this diversity of thought supports leaders to embrace the complexity and diverse nature of employees' perspectives and experiences and how they influence the system through their interconnection. Furthermore, embracing complexity in this way helps to promote a culture of learning, collaboration and resilience within the organisation. It empowers employees at all levels to contribute their knowledge, creating a shared sense of ownership and commitment (Senge, 1990).

Drawing out diversity of experience and thought is helpful, but once that has occurred, the leader also needs to be able to notice the patterns underlying connections and trends if they want to understand the 'cause' of issues within a complex system. They need to be willing to seek out and identify recurring themes, causal loops and systemic connections between parts of the system. This is a skill that comes more readily to some, and one leaders can develop through engaging in data-driven decision making. Leaders can foster a culture of collecting and analysing relevant data from various sources to understand an issue rather than relying solely on anecdotal evidence or assumptions. They must also be willing to seek external perspectives on an issue, engaging with external stakeholders, consultants or experts who can potentially identify patterns that may be overlooked by those immersed in day-to-day operations. They can engage in 'systems mapping' exercises (I have included an example in the activities at the end of this chapter) to actively map out, in a visual way, the interconnectedness of various systems. The more they can develop the skills of identifying patterns in a system or a 'problem', the more able they will be to address the root causes effectively, anticipate potential future issues and develop sustainable solutions to complex challenges.

Finally, I return to the importance of the resilience mindset – one characterised by a willingness to adapt and flex in the face of adversity of challenge. A resistance to change and adaptability is not going to serve you well as a systemic leader; instead you need to be willing to embrace the reality of the VUCA world we live in and be willing to explore new opportunities and experimentation. You also need to be engaged in developing the resilience and agility of the system as a whole, breaking down silos, encouraging cross-functional collaboration and empowering teams to make decisions quickly in response to changing demands and market conditions. For the ego-driven leaders we discussed in the chapter 8, this will be something they find particularly difficult because it requires leading from the position of 'Us' rather than 'Me'. By developing the competencies of a systemic leadership perspective, leaders can learn to navigate complexity, foster inclusivity, build trust, communicate effectively, balance short-term goals with long-term impact and solve issues at the root to drive sustainable organisational success.

Encouraging Systemic Leadership in the System

The system itself can also engage in developing systemic leadership skills in its leaders through organising workshops or training programmes that focus on developing systems thinking skills. These sessions should equip leaders with the tools to analyse complex systems by taking a helicopter view and appreciate in the influential relationships between elements of the system. You can back these programmes up by encouraging cross-functional collaboration projects. This will support leaders to gain a holistic view of the organisation by directly interacting with different elements of it. Following these projects up with action-learning assignments where leaders can reflect on and apply their learning will help ensure that the ideas stick. A training programmes or no follow-up is one of the biggest wastes or resources in my opinion. If you want people to retain knowledge, you need to get them to use it. Engaging in regular cross-functional projects and following these up with action learning can help develop the learning culture that is essential for the understanding and adaptability required of a systemic leader (Senge, 1990).

You can embed systemic leadership practices across the whole system by championing the leaders who are already engaged in it. Recognise and reward behaviours that align with systemic leadership principles, such as collaboration, innovation and results orientation. Celebrate successes and acknowledge contributions that contribute to the organisation's overall success. Set up mentorship programmes and coaching where systemic champions can share knowledge and practise with less agile leaders. You can help these qualities to grow by nurturing them when they develop or where they are already present. Finally, if you want your leaders to be able to hold the 'bigger picture' in mind and make decisions for the future benefit of the company, you need to have a clear vision for what that future looks like. Define a clear organisational vision and values that incorporate systemic thinking, and communicate these values regularly if you want them to guide decision making and behaviour at all levels of the organisation.

As with any cultural shift, the shift to systemic leadership is likely to be met with resistance. Remember humans don't like change. You can help overcome this through education and training initiatives, such as webinars, workshops or online programmes to develop understanding and get collective buy-in. Through these you can identify people's beliefs and concerns and highlight the benefits for achieving organisational goals. If you want people to support the process of change, then involve them in the process as much as possible.

Don't try to shift the whole system all at once. Remember that systems shift more readily from the edges. Engage in pilot projects or 'experiments' to demonstrate the effectiveness of systemic leadership principles in practice. Then, gather data, measure outcomes and evaluate the impact of pilot initiatives before scaling up to broader organisational change efforts. Use pilot projects as learning opportunities to test assumptions, refine strategies and build momentum for change, reducing resistance and increasing stakeholder buy-in over time. Most importantly, recognise that cultural transformation takes time and requires sustained effort. Foster a supportive environment where employees feel empowered to embrace change, challenge the status quo and contribute to organisational success. Provide them with the 'How' and also the 'Why' to help them align your goals with theirs. By adopting these strategies, anticipating and overcoming challenges, organisations can begin to shift to systemic leadership culture that promotes collaboration, innovation and sustainable growth.

Systemic leadership represents a paradigm shift, going beyond traditional hierarchical models of leadership and embracing a holistic perspective that acknowledges the interconnectedness of all elements within an organisation. At its essence, systemic leadership entails a shift from linear thinking to systems thinking, where leaders recognise the complex web of relationships and interdependencies that influence organisational outcomes. If you are looking for a leadership approach that helps you navigate the VUCA age, drive positive change, create a culture where employees continue to thrive and everyone is oriented towards the same goals, then systemic leadership is it.

Leading Systems: Activities for Developing Systemic Leadership

Few systemic leaders are born; most are made. The following are a range of activities to begin to build systemic leadership qualities in your leaders.

Team Activity	*Mapping the Employee Engagement System*
Objective	The System Mapping activity is designed to help participants understand the interconnected factors influencing employee engagement within the organisation and develop a systemic perspective on addressing this complex issue.
Duration	1–1.5 hours
Materials Needed	Meeting room or designated space conducive to open discussion, flip chart paper or whiteboard, markers, sticky notes, and pens.

Here's how it works:

Introduction (10 minutes): Introduce the purpose of the workshop. Set expectations of open communication, respect and confidentiality. Outline the agenda and any other ground rules.

Divide participants into small groups of 4–6 people, ensuring a mix of individuals from different departments or functions within the organisation

Explain the context of the exercise. Remind participants that Employee Engagement is necessary for organisational performance, productivity and retention. However, employee engagement is influenced by various interconnected factors within the organisational system.

Group Work (20 Minutes): Ask each group to identify and list as many factors as possible that might contribute to or influence employee engagement within the organisation. These could include leadership and management practices, organisational culture, work processes and systems, development opportunities, compensation and benefits, work-life balance and wellbeing initiatives, physical work environment and communication and feedback mechanisms.

System Mapping (20 minutes): Once the groups have generated a comprehensive list of factors, distribute sticky notes of different colours and ask them to write each factor on a separate sticky note. Instruct the groups to create a system map by arranging the sticky notes on the large paper or whiteboard, grouping related factors together and using arrows or lines to illustrate the connections and relationships between different factors.

Encourage participants to discuss and debate the interconnections, potential feedback loops and ripple effects within the system. They should consider how changes in one factor might influence or be influenced by other factors across different departments or functions.

Group Discussion (15 minutes): After the mapping exercise, facilitate a group discussion. Ask participants to share their insights, observations and learnings from the activity. Encourage them to reflect on the complexity and interconnectedness of the employee engagement system and how a systemic perspective can inform more effective and sustainable strategies.

(Continued)

Debrief & Action Planning (10 minutes): Reflecting one key learning points from the session encourage participants to identify actions they can take away to begin to influence the system to improve employee engagement. This activity promotes several systemic leadership qualities:

1. Systems thinking: Participants practice visualizing and understanding the interconnections and relationships within the complex system of employee engagement.
2. Holistic perspective: By considering various factors across different domains (leadership, culture, processes, development, environment, etc.), participants develop a more comprehensive understanding of the issue.
3. Cross-functional collaboration: Having participants from different departments work together promotes cross-functional collaboration and helps break down silos within the organisation.
4. Embracing complexity: The activity encourages participants to acknowledge and embrace the complexity of employee engagement rather than seeking oversimplified solutions.
5. Continuous learning and adaptation: The discussion following the mapping exercise can highlight the need for continuous learning, adapting strategies and incorporating feedback loops to address the evolving dynamics of the employee engagement system.

This activity can be tailored to different organisational contexts by adjusting the central issue or system being mapped (e.g., customer experience, innovation, operational efficiency). The hands-on nature of the activity and the group discussions promote collaboration, systems thinking, embracing complexity and a deeper appreciation for systemic leadership principles in an organisational setting.

Team Activity	*Systemic Leadership Simulation*
Objective	The objective of this activity is to assess participants' understanding of systemic leadership principles and their ability to apply them in various organisational contexts. Through a simulation exercise, participants will have the opportunity to actively practice systemic thinking, collaboration and decision making while addressing complex challenges.
Duration	2–2.5 hours
Materials Needed	Meeting room or designated space conducive to open discussion, flip chart paper or whiteboard, markers, sticky notes, pens, simulation scenario (created by the facilitator).

Here's how it works:

Introduction (10 minutes): Introduce the purpose of the workshop. Set expectations of open communication, respect and confidentiality. Outline the agenda and any other ground rules.

Begin by introducing the concept of systemic leadership and its importance in organisational success. Provide an overview of the activity and its objectives.

(Continued)

Simulation Setup (10 minutes): Divide participants into small groups, ideally representing diverse functions or departments within the organisation. Distribute the simulation scenario, which presents a complex organisational challenge that requires systemic leadership to address effectively.

Simulation Exercise (60 minutes): Instruct each group to read the scenario, and identify the key stakeholders, challenges, and opportunities presented. Encourage participants to apply systemic leadership principles, such as open-mindedness, attention to diversity, value of relationships, effective communication, result orientation and co-creation of structures and strategies, as they work through the simulation. Provide guidance and support as needed, but allow groups to lead their own discussions and decision-making processes.

Debrief & Reflection (30 minutes): Reconvene the groups and facilitate a debrief session to discuss their experiences during the simulation. Encourage participants to reflect on how they applied systemic leadership principles in their decision-making processes and interactions with stakeholders. Facilitate a discussion on the challenges encountered and lessons learned, highlighting areas for improvement and further development of systemic leadership skills.

Action Planning (20 minutes): Guide participants in developing action plans for applying systemic leadership principles in their day-to-day roles and responsibilities.

Encourage participants to identify specific actions they can take to enhance their systemic leadership capabilities, such as seeking feedback, building relationships with cross-functional teams or attending training sessions.

References

Gharajedaghi, J. (2011). *Systems Thinking: Managing Chaos and Complexity: A Platform for Designing Business Architecture* (3rd ed.). Morgan Kaufmann.

Liker, J. K. (2004). *The Toyota way: 14 management principles from the world's greatest manufacturer.* McGraw-Hill.

Meadows, D. H. (2008). *Thinking in Systems: A Primer.* Chelsea Green Publishing.

Messersmith, D. H. (1993). *Lincoln memorial lighting and midge study.* Unpublished report prepared for the National Park Service. CX-2000-1-0014.

Senge, P. M. (1990). *The fifth discipline: The art and practice of the learning organisation.* Doubleday/Currency.

Wheatley, M. J. (2006). *Leadership and the new science: Discovering order in a chaotic world.* Berrett-Koehler Publishers.

12 Culture as a KPI: Why Investing in a Sustainable Wellbeing Culture Makes Business Sense

Happy Bees Work Harder

Investing in consciously cultivating a positive workplace culture, and specifically a wellbeing-oriented culture, isn't just about creating a feel-good atmosphere; it's about building a sustainable culture where employees continue to thrive and businesses flourish. When employees feel valued, supported and empowered to prioritise their wellbeing and leaders are equipped to see the system as a whole, they become the backbone of a resilient and high-performing organisation.

First and foremost, prioritising employee wellbeing is essential for cultivating a sustainable workplace culture. By investing in initiatives that promote physical health, mental wellbeing and self-care, organisations create an environment where employees feel cared for. This, in turn, encourages a sense of belonging and loyalty, resulting in higher engagement, productivity and retention. As the saying goes, "happy bees work harder", and when employees feel happy and fulfilled, they are more likely to go above and beyond to contribute to the success of the organisation. Similarly encouraging your leaders to adopt a systemic leadership perspective can help them nurture this positive environment and activity identify problems and sources of disruption and seek solutions that are sustainable in the longer term and work within the culture rather than against it.

As well as the obvious moral and ethical motivations, a focus on developing a healthy culture makes strategic business sense. Research consistently shows that organisations with a strong emphasis on employee wellbeing experienced reduced absenteeism, higher levels of productivity, improved organisational resilience and adaptability, enhanced reputation and greater alignment of organisational values and vision. When employees exist in a culture that supports their wellbeing and nurtures their need for purpose and meaning, they are more present, focused and productive at work. This not only drives organisational performance but also leads to significant cost savings and improved financial outcomes (Baicker et al., 2010).

Additionally, investing in culture development strategies helps organisations attract and retain top talent. In an era where employee experience

DOI: 10.4324/9781003407577-12

and workplace culture play a crucial role in recruitment and retention, organisations that prioritise holistic wellbeing have a distinct advantage. Employees are more likely to choose and stay with employers who prioritise their wellbeing and offer a supportive and inclusive work environment with genuine opportunities for growth and personal development.

Culture as a KPI: What's Good for the Bees …

It's time that you viewed culture as a key performance indicator (KPI). The traditional metrics of revenue and profitability are far too linear to understand the health of a System. "What is good for the bees is good for the hive,", understanding that the interconnectedness between culture, employee wellbeing and organisational success is the real metric that matters, and this is something we have known for a while.

Investing in employee wellbeing leads to higher levels of engagement, motivation and productivity. Research by Gallup, a multinational analytics advisory company, highlighted that engaged teams are 21% more profitable, with 17% higher productivity compared with disengaged teams (Gallup, 2022). If you want that in 'real money', a 2020 study by Deloitte found that for every dollar that companies invested in mental health interventions, they saw a return of $4 in terms of productivity (Deloitte, 2020).

A healthy, sustainable culture also improves resilience and retention. Staff are more likely to stay at an organisation where they feel fulfilled and that their needs are being met. A recent report produced by The Work Institute found that organisations with a positive culture and engaged employees had 40% lower annual turnover than those without. They also noted that 'poor workplace culture' was cited as a reason for leaving by 22% of employees who voluntarily quit their jobs (Work Institute, 2022).

This has an impact on reputation also. A positive workplace culture helps you not only retain talent but also attract talent. Looking back at the previous example of Patagonia's commitment to employee wellbeing and environmental sustainability (Chapter 8), their approach has boosted employee morale as well as enhancing the company's brand reputation, improving talent acquisition and customer loyalty.

Investing in culture also helps build an agile, resilient workforce who are able to adapt to change and manage stress. Wellbeing-oriented cultures that are aimed at reducing stress, promoting work-life balance and fostering mental health awareness contribute to lower absenteeism and presenteeism rates. A 2022 study, published in the *Journal of Occupational and Environmental Medicine*, noted that organisations that implemented a stress management programme saw an average 33% decrease in absenteeism and a 27% reduction in presenteeism rates compared to control groups after one year. The researchers estimated cost savings to be about $5,000 per employee per year after accounting for programme costs (Wilson et al., 2022).

Re-imagining culture as a KPI reflects an organisation's understanding of the significant impact a healthy workplace culture has on the long-term sustainability and success of the system. It also sends a powerful message that employee wellbeing is not just an afterthought but also a fundamental aspect of the organisation's values, identity and purpose.

'Culture as a KPI' reflects a holistic approach to organisational performance, by prioritising employee wellbeing, engagement, retention, resilience and brand reputation, ultimately driving sustainable growth. As leaders, it is imperative to recognise that "what is good for the bees is good for the hive" and invest in consciously cultivating the systemic culture of the organisation for the benefit of everyone in the system.

Transitioning from Why to How

Hopefully by now you have understood the 'Why' of cultivating culture and orientating it towards wellbeing. Now we will look at the 'How'. Remember, "a change in behaviour is far more likely to lead to a change in awareness, than a change in awareness will lead to a change in behaviour". It's time to more from 'theory and strategy' to 'tactics and actions', and how to design, implement and evaluate culture change strategies for organisations of any size. Let's explore both short-term and long-term approaches that can be implemented today.

Step 1: Understand the Culture You Have

Start by actively understanding your culture as it stands. Conduct an anonymous employee survey, and send it out across the entire organisation to gather feedback. To get a comprehensive overview of your organisation's culture, you can include questions like these under the following headings:

General Culture Assessment

1. *How would you describe the current culture of our organisation in a few words?*
2. *On a scale of 1–10, how satisfied are you with the current organisational culture?*
3. *What aspects of our company culture do you value the most?*
4. *What aspects of our company culture do you feel need improvement?*

Communication and Transparency

5. *Do you feel that communication from leadership is clear and transparent? Why or why not?*
6. *How frequently do you receive updates on company goals and performance?*
7. *How comfortable do you feel providing feedback to your manager or senior leadership?*

Values and Mission

8. *Do you understand and align with the company's mission and values?*
9. *How often do you see the company's values reflected in everyday work and decision making?*
10. *Are there any values you believe should be added or emphasised more?*

Employee Engagement and Inclusion

11. *Do you feel valued and recognised for your contributions to the company?*
12. *How included do you feel in the decision-making processes that affect your work?*
13. *How well does the company support diversity and inclusion in the workplace?*

Work Environment and Collaboration

14. *How would you rate the level of collaboration between teams/departments?*
15. *Do you feel that you have the tools and resources necessary to do your job effectively?*
16. *How supportive is the work environment in terms of work–life balance?*

Leadership and Management

17. *How would you rate the effectiveness of your immediate supervisor/manager?*
18. *Do you believe that leadership is approachable and open to new ideas?*
19. *How well does leadership handle conflict and resolve issues?*

Professional Development

20. *How satisfied are you with the opportunities for professional growth and development within the company?*
21. *Do you feel that the company invests adequately in employee training and development?*

Change Readiness

22. *How open are you to changes in our company culture?*
23. *What concerns do you have about potential changes to the company culture?*
24. *What support do you think is necessary to successfully implement culture changes?*

Open-Ended Questions

25. *What do you think is the biggest cultural challenge our company faces?*
26. *Can you provide an example of a time when you felt the company culture positively impacted your work?*
27. *Can you provide an example of a time when you felt the company culture negatively impacted your work?*
28. *What one change would you make to improve the company culture?*

Once you have collected the data, analyse it and combine your findings with those from exit interviews, performance reviews and any other employee engagement surveys you have conducted. Draw out key findings and begin to paint the picture of what your culture looks like.

You can also use a strategic planning tool such as a SWOT (Strengths, Weaknesses, Opportunities, Threats) analysis to understand internal and external factors, including those impacting the culture (Dyson, 2004). This analysis tool is widely used for future planning and can help you understand your current cultural position and develop strategies for improvement. I have detailed the eight-phase process for a detailed SWOT analysis of culture at the end of this chapter.

Step 2: Define the Desired Culture

Engage in a visioning exercise to identify the organisation's purpose, values and desired behaviours. What is the culture you are looking to cultivate, and what is the culture your organisation needs? You can start by asking yourself these questions.

1. *What are our core values and beliefs as an organisation?*

   • *What principles should guide our decision making and behaviours?*
   • *What values do we want to be known for and recognised by?*

2. *What is our purpose and mission?*

   • *Why does our organisation exist, beyond just making a profit?*
   • *What positive impact do we want to have on our stakeholders (customers, employees, community, etc.)?*

3. *What behaviours and attitudes do we want to encourage and celebrate?*

   • *What does collaboration, teamwork, and open communication look like in our ideal culture?*
   • *How do we want to approach problem-solving, innovation and continuous improvement?*

4. *How do we want to treat and support our employees?*

   • *What is our desired work environment and employee experience?*
   • *How do we foster trust, respect and a sense of belonging among our workforces?*

5. *What is our desired leadership style and approach?*

   • *How do we want our leaders to interact with and empower their teams?*
   • *What leadership qualities and competencies are essential for our desired culture?*

6. *How do we want to be perceived by our customers and external stakeholders?*

   • *What kind of customer service and experience do we want to provide?*
   • *What ethical standards and social responsibility practices are important to us?*

7. **What are our competitive advantages and unique strengths?**

- *What cultural elements can help us differentiate ourselves in the market?*
- *How can our culture support our strategic goals and vision for the future?*

8. **What cultural elements from our past do we want to preserve or evolve?**

- *What positive aspects of our current culture should we build upon?*
- *What aspects of our current culture may need to change to align with our desired future state?*

9. **How can we create a culture that attracts and retains top talent?**

- *What kind of work environment and development opportunities are important to our ideal candidates?*
- *How can our culture support employee engagement, motivation and retention?*

10. **What cultural shifts are necessary to adapt to changing market conditions or industry trends?**

- *How can our culture support innovation, agility and resilience in the face of change?*
- *What cultural elements can help us stay competitive and relevant in the long term?*

Refine your findings by involving a diverse group of employees from different levels, departments and backgrounds across the organisation. Conduct focus groups or workshops to gather a range of input and alignment. Then once you think you have it, write it up and run the groups again to make sure you've got it right.

Step 3: Gain Leadership Commitment

Securing leadership commitment to a culture change programme is essential for its success. The first step is to ensure that top leadership, including executives and senior managers, is fully committed to the initiative. This begins with engaging them early in the process, presenting compelling data and illustrating the direct benefits of culture change to the organisation's overall performance and goals. By aligning the culture change programme with the overall business strategy, you can demonstrate its importance and relevance, making it clear that their involvement is crucial.

Once commitment is established, it's important to provide training and coaching to cultivate the necessary systemic leadership skills to support a culture shift. Change management training is vital, as it helps leaders understand the complexities of organisational change and how to effectively manage resistance. Workshops on communication skills are equally important. These sessions should focus on ensuring leaders can convey the vision and necessary changes clearly and persuasively. Additionally, offering personalised leadership coaching can help leaders develop their systemic skills, model desired behaviours and become more wellbeing oriented.

To further solidify commitment, establish a culture change steering committee or taskforce. This group should be led by senior leaders and include representatives from various departments. Regular meetings should be held to discuss progress, address challenges and ensure that the initiative stays on track. Empowering the taskforce with decision-making authority will enable them to implement necessary changes effectively. Developing a leadership accountability plan is another crucial step. This plan should clearly outline the roles and responsibilities of each leader in supporting the culture change. It should also set specific expectations for behaviours and actions that align with the desired culture. Regularly monitoring progress and providing feedback will help ensure that leaders remain committed and on course.

Open communication from leaders is essential throughout the culture change process. Leaders should communicate clearly about the reasons for the change, explaining the benefits for the organisation and its employees. Make sure to sharing success stories and milestones as well as examples of how the culture change has positively impacted other organisations or internal teams can be very persuasive. Moreover, creating opportunities for two-way dialogue allows employees to ask questions, express concerns and feel involved in the process.

Your leaders are the cultural curators in your system, securing their commitment is essential to drive a successful culture change programme. Leaders must be fully engaged, well trained, accountable and transparent in their communications to foster a positive and effective cultural transformation.

Step 4: Develop a Communication Plan

Now that you have an idea of the culture you want to cultivate and have got buy-in from your leaders, you need to consider how you will communicate the process of change across the organisation. You need to develop a comprehensive communication that addresses all stakeholders, including employees, customers and partners, adjusting your communication to meet the needs of different groups.

Using various channels is crucial for effective communication. Employ a mix of emails, town halls, newsletters, the intranet and social media to ensure that the message reaches all audiences. Different people consume information in different ways.

Clear and consistent messaging is essential. The messages should articulate the desired culture and the reasons for the change. This clarity helps to minimise confusion and resistance. It is important to explain not only what changes are being made but also why they are necessary and how they will benefit the organisation and its stakeholders. Align key messages with your vision and your 'Why'. For example:

Vision Statement: "Our goal is to create a culture that prioritises innovation, collaboration and inclusivity".
Rationale for Change: "Adapting our culture is essential to stay competitive, improve employee satisfaction and better serve our customers".

Benefits: "A positive culture will enhance job satisfaction, increase productivity, attract top talent and drive business success".

In addition, clearly articulate the new values and behaviours you expect.

Collaboration

- **Value Statement:** "We believe that collaboration is key to our success. By working together, we can achieve more than we ever could alone".
- **Expected Behaviour:** "We expect every team member to actively participate in group projects, share knowledge freely and support their colleagues in achieving common goals".

Innovation

- **Value Statement:** "Innovation drives our progress. We are committed to fostering a culture where creativity and new ideas are encouraged and valued".
- **Expected Behaviour:** "We encourage everyone to think outside the box, propose new solutions and embrace change as a constant opportunity for improvement".

Inclusivity

- **Value Statement:** "Inclusivity is at the heart of our culture. We are dedicated to creating a workplace where everyone feels valued, respected and has equal opportunities to contribute and grow".
- **Expected Behaviour:** "We expect all employees to treat each other with respect, celebrate diverse perspectives and ensure that every voice is heard and considered".

Make sure that your communication strategy includes space for open dialogue. Provide opportunities for employees to ask questions and offer feedback. This can be done through Q&A sessions, feedback forms or informal meetings. When employees feel heard, they are more likely to buy into the change. It also helps to identify and address any concerns early in the process.

Step 5: Align Policies and Practices to Embed the New Culture

The successful implementation of your culture change will require a framework of policies and practices that align with your new cultural values. This alignment begins with a thorough review and revision of HR policies, procedures and systems. You can adjust recruitment processes to prioritise candidates who demonstrate your desired cultural values. You will also need to update your performance management processes so that they reflect and reinforce these values, focusing on behaviours and outcomes that align with the new culture.

Compensation and benefits programmes should be realigned to support and reward the desired cultural traits. For instance, you might introduce incentives for teamwork, creativity and inclusivity. By linking rewards and recognition practices to the new cultural values, you reinforce the importance of these behaviours and motivate employees to adopt them.

Training and resources are critical in helping employees adapt to the new culture. Providing comprehensive training programmes that focus on developing skills and behaviours consistent with the desired culture will support employees in making the transition. These programmes might include workshops on effective collaboration, innovative thinking and fostering an inclusive work environment. Continuous learning opportunities should also be available to ensure ongoing development and reinforcement of these behaviours.

It is equally important to identify and eliminate practices or processes that may be counterproductive or misaligned with the new culture. Conduct a thorough audit of existing practices to pinpoint areas that conflict with the desired cultural attributes. For example, if existing processes discourage open communication or collaborative work, they must be restructured to align with the new cultural goals.

By aligning policies and practices with the desired culture, you create an environment where the new values are consistently supported and reinforced. This alignment not only facilitates the culture change but also ensures that the new culture is sustainable and deeply embedded within the organisation.

Step 6: Measure and Sustain It

Measurement is key to sustaining a culture change programme, noticing what is working, what is not and what needs adjusting. It's crucial to establish clear metrics and KPIs to track implementation. These metrics may include employee engagement scores, turnover rates, customer satisfaction ratings and operational performance indicators. By regularly monitoring these metrics, you can continually assess progress.

In addition to quantitative data, like that collected in standard engagement surveys, gather qualitative feedback on employees' perceptions of the new cultural values, their level of engagement and any barriers they may be facing in adopting the desired behaviours. Analysing both sets of data, alongside operational metrics, will provide a deeper understanding of the culture change programmes impact by identifying trends across the data sets and assessing how the programmes is influencing various aspects of the business.

Continuous communication is essential for sustaining momentum and engagement throughout the culture change initiative. Leaders should regularly communicate progress updates, successes, early 'wins' and key learnings to employees at all levels of the organisation. This transparency helps to reinforce the importance of the culture change, maintain employee buy-in and demonstrate leadership commitment to the initiative.

Finally, you need to be prepared to adjust and refine your strategy based on feedback and results. If certain aspects of the culture change programme are not yielding the desired outcomes, it may be time to revisiting and revising your approach. By remaining flexible and responsive to feedback, you can ensure that your culture change efforts remain relevant and effective in the long term. Remember, culture change is an ongoing journey, and it requires patience, persistence and a long-term commitment from the entire organisation. A solid foundation, clear communication, continuous reinforcement, measurement and adjustment are crucial to sustaining the desired culture over time.

Culture as a KPI: Analysis and Communication in Culture Change

Before we shift to the culture we want, we need to understand the culture we have. Alongside surveys and feedback, you can conduct a SWOT analysis to examine the Strengths, Weaknesses, Opportunities and Threats. Then, once we have used this to formulate a model for our design culture and the programme of change to implement it, we need a strategy to communicate that change across the system. I have provided an outline of how this can be done in the context of SWOT culture assessment and a comprehensive communication strategy. Use these as examples to inform your own approach:

Activity 1	*SWOT Analysis for Culture Assessment*
Phase 1: Preparation	1. **Form a Team:** Assemble a diverse group of employees from different departments and levels within the organisation. This team should include senior leaders, middle managers and frontline employees to ensure a variety of perspectives. 2. **Gather Data:** Collect existing data from employee surveys, performance reviews, feedback sessions, exit interviews and any other relevant sources. This data will provide a foundation for your analysis.
Phase 2: Identify Strengths	'Strengths' describe the internal attributes of the organisation that are beneficial to achieving the company's objectives. • **Questions to Consider:** • What are the positive aspects of our current culture? • What unique cultural traits give us a competitive edge? • Which values or behaviours are most praised and appreciated by employees? • What aspects of our culture attract and retain top talent? • **Example Strengths:** • *Strong sense of teamwork and collaboration* • *High levels of employee engagement* • *Transparent and effective communication from leadership* • *Robust training and development programmes*

(Continued)

Phase 4: Identify Weaknesses	'Weaknesses' are the internal factors that are harmful to achieving the company's objectives.
	• **Questions to Consider:**
	  • What cultural aspects are frequently criticized by employees?
	  • Where do we experience the most significant communication breakdowns?
	  • What behaviours or values are misaligned with our goals?
	  • In what areas do we fail to support our employees effectively?
	• **Example Weaknesses:**
	  • *Lack of diversity and inclusion*
	  • *Poor work-life balance*
	  • *Insufficient recognition and reward systems*
	  • *Resistance to change among certain employee groups*

Phase 4: Identify Opportunities	'Opportunities' are the external factors that can potentially be leveraged to the organisation's advantage.
	• **Questions to Consider:**
	  • What cultural trends are emerging in our industry that we can adopt?
	  • Are there technological advancements that can enhance our culture?
	  • What external resources (consultants, training programmes) can we utilize to improve our culture?
	  • How can changes in the market or economy positively impact our culture?
	• **Example Opportunities:**
	  • *Increasing remote work capabilities*
	  • *Partnering with diversity and inclusion organisations*
	  • *Leveraging new communication tools to improve transparency*
	  • *Capitalising on industry trends to foster innovation*

Phase 5: Identify Threats	'Threats' are the external factors that could negatively impact the organisation's culture.
	• **Questions to Consider:**
	  • What external challenges could negatively impact our culture?
	  • Are there industry shifts that threaten our current cultural practices?
	  • How do our competitors' cultures pose a threat to us?
	  • What external socio-economic changes could impact employee morale and engagement?
	• **Example Threats:**
	  • *Economic downturn affecting job security*
	  • *Competitive pressures leading to high turnover*
	  • *Legal or regulatory changes impacting workplace policies*
	  • *Technological disruptions making current practices obsolete*

(Continued)

Phase 6: Analyse and Prioritise	**Group Discussion:** Conduct a session where team members present their findings. Encourage open discussion and debate to ensure a comprehensive understanding of each point. 1. **Prioritise Issues:** Use tools like voting or ranking to prioritise the most critical strengths, weaknesses, opportunities and threats. 2. **Develop Strategies:** Based on the prioritised list above, develop strategies to respond to each.
Phase 7: Action Plan	1. **Set Objectives:** Define clear, actionable objectives for cultural change based on the SWOT analysis. 2. **Assign Responsibilities:** Designate team members or departments responsible for each initiative. 3. **Create a Timeline:** Develop a realistic timeline for implementing each action item. 4. **Monitor and Adjust:** Establish metrics to measure and monitor progress and be prepared to adjust strategies as needed.
Phase 8: Communicate & Engage	1. **Share Results:** Communicate the results of the SWOT analysis and the planned actions to all employees to ensure transparency. 2. **Engage Employees:** Involve employees in the implementation process through feedback loops, focus groups and regular updates.

Activity 2	*Developing A Comprehensive Communication Strategy For A Culture Change Initiative*
Introduction	• **Purpose:** Define the purpose of the communication strategy, which is to support the culture change programme by ensuring all stakeholders are informed, engaged and aligned with the new cultural values. • **Goals:** Enhance understanding of the change, foster acceptance, promote active participation and sustain momentum throughout the change process.
Stakeholder Identification And Analysis	• **Identify Stakeholders:** Employees, managers, executives, customers, partners, and other relevant stakeholders. • **Stakeholder Needs:** Assess the specific needs, concerns and preferred communication channels for each stakeholder group.
Key Messages	• **Vision Statement:** Include a vision statement such as – "Our goal is to create a culture that prioritises innovation, collaboration and inclusivity". • **Rationale for Change:** Include you 'Why' for the culture change programmes – "Adapting our culture is essential to stay competitive, improve employee satisfaction and better serve our customers".

(Continued)

	• **Benefits:** Help people understand the benefits – "A positive culture will enhance job satisfaction, increase productivity, attract top talent and drive business success". • **Values and Behaviours:** Clearly articulate the new values and expected behaviours of the new culture, e.g., teamwork, transparency and continuous improvement.
Communication Channels and Tools	• **Emails:** Regular updates and messages from leadership to keep everyone informed. • **Town Halls:** Quarterly town hall meetings led by senior leaders to discuss progress and answer questions. • **Newsletters:** Monthly newsletters with articles, interviews and updates related to the culture change. • **Intranet:** A dedicated section on the company intranet with resources, FAQs, forums for discussion and a feedback portal. • **Social Media:** Use internal social media platforms (e.g., Slack, Yammer) for real-time updates and engagement. • **Workshops and Training Sessions:** Interactive sessions to educate employees on the new cultural values and behaviours.
Communication Timeline	• **Start:** Launch the communication strategy with a major announcement from the CEO, outlining the vision and goals. • **Initial Phase (Months 1–3):** Frequent updates and intensive engagement activities, including weekly emails and biweekly town halls. • **Middle Phase (Months 4–9):** Maintain momentum with monthly newsletters, quarterly town halls and regular training sessions. • **Ongoing Phase (Months 10+):** Ensure sustainability with periodic updates, annual culture reviews and continuous engagement activities.
Engagement and Dialogue Mechanisms	• **Q&A Sessions:** Regularly scheduled sessions where employees can ask questions directly to leaders. • **Feedback Mechanisms:** Online feedback forms, suggestion boxes and focus groups to gather employee input. • **Listening Tours:** Leaders visit various departments to listen to employee concerns and suggestions. • **Surveys:** Conduct regular surveys to measure employee sentiment and gather feedback on the culture change process.
Celebrating Milestones and Successes	• **Milestone Celebrations:** Acknowledge key milestones in newsletters and town halls. • **Employee Recognition:** Highlight and reward employees who demonstrate your new cultural values. • **Success Stories:** Share stories of teams or individuals who have successfully embraced the new culture.

(Continued)

Leadership Involvement and Visibility	• **Leadership Messages:** Frequent messages from senior leaders to reinforce commitment and provide updates. • **Role Modelling:** Leaders actively demonstrate the desired behaviours and values. • **Visibility:** Leaders maintain a visible presence in communication channels and engagement activities.
Training and Development	• **Change Management Training:** Provide training for leaders on managing change, effective communication and leading by example. • **Culture Workshops:** Conduct workshops to help employees understand and adopt the new cultural values. • **Ongoing Development:** Offer continuous learning opportunities to reinforce the desired culture.
Monitoring and Evaluation	• **Metrics and KPIs:** Establish metrics to track the progress of the culture change (e.g., employee engagement scores, turnover rates, productivity). • **Regular Reviews:** Conduct quarterly reviews to assess the effectiveness of the communication strategy and make necessary adjustments. • **Feedback Analysis:** Regularly analyse feedback from employees and other stakeholders to identify areas for improvement.

References

Baicker, K., Cutler, D., & Song, Z. (2010). Workplace wellness programs can generate savings. *Health Affairs, 29*(2), 304–311.

Deloitte. (2020). *Mental Health and Employers: Refreshing the Case for Investment.* Deloitte LLP. Available at: www2.deloitte.com/uk/en/pages/consulting/articles/mental-health-and-employers-refreshing-the-case-for-investment.html (Accessed: 11 April, 2024).

Dyson, R. G. (2004). Strategic development and SWOT analysis at the University of Warwick. *European Journal of Operational Research, 152*(3), 631–640.

Gallup. (2022). *State of the Global Workplace: 2022 report.* Gallup Press. Available at: gallup.com/workplace/349484/state-of-the-global-workplace-2022-report.aspx (Accessed: 11 April, 2024).

Wilson, M., Frank, F., & Jamieson, S. (2022). The impact of a workplace stress management program on absenteeism, presenteeism, and productivity in North American workers. *Journal of Occupational and Environmental Medicine, 64*(5), 365–376.

Work Institute. (2022) *How employers caused the great resignation.* Available at: info.workinstitute.com/hubfs/2022%20Retention%20Report/2022%20Retention%20Report%20-%20Work%20Institute.pdf (Accessed: 11 April, 2024).

13 Recap and Reflection

As you come to the end of this book, I want to reflect back on the journey so far through the complex ecosystem of workplace culture, what a healthy sustainable culture looks like and how we can consciously build one. By developing your self-awareness and understanding the importance of employee needs, you can begin to consider how both sit within the context of a sustainable system.

The journey always starts with greater self-awareness, reflecting on the impact of your beliefs, thoughts and feelings on your behaviour and perceptions. Once we have a better understanding of ourselves, we can begin to extend that understanding out to other people. It is greater self-awareness that is necessary to practise the dynamic skills of Emotional, Social and Relational Intelligence, facilitating the level of conscious communication that is the foundation for developing positive workplace relationships. It is this same self-awareness that will help you address the unconscious biases that are getting in the way of creating an inclusive environment where everyone feels valued and respected.

We considered the perspective of the 'other', in this case the people who make up your organisational system. We considered the importance of core human needs that have remained relatively constant over millennia, alongside implicit psychological contracts and how they influence the underlying motivations and expectations that drive employee behaviour. We noted how when people feel their needs are being met, within an organisational context, it leads to greater engagement and connection, increasing a felt sense of safety and belonging. Alongside this, we identified the significance of values in shaping personal beliefs and behaviours as well as organisational culture, underscoring the need for alignment between our stated values and our enacted behaviours. Walking the walk as much as we talk the talk.

We explored how change and even crises are inevitable consequences of a complex organisational system in a changing work environment. The only constant in life is change. As such, engaging in strategies to build both personal and organisational resilience can support individuals and organisations to adapt to change and navigate challenges with less stress, engaging proactively rather than reactively for greater sustainability and less disruption.

DOI: 10.4324/9781003407577-13

We ended our journey by taking a step back to consider the system, recognising the interconnectedness of individuals and the broader community, emphasising the need for compassion, empathy and social responsibility in our interpersonal interactions as well as in our organisational behaviours. This serves to develop the necessary psychological safety required in a healthy culture, where individuals feel safe to express themselves, take risks and innovate without fear of judgement or reprisal. Holding this approach within a framework of systemic leadership helps recognise the role of leaders in cultivating sustainable cultures as well as reinforcing the healthy behaviours and positive environment that encourage the organisation to thrive in the long term.

Hopefully, it has become evident that cultivating a culture oriented towards employee wellbeing is essential for any organisation to be functional in a truly sustainable way; indeed, it's likely a big part of the reason your organisation is not performing as you would like right now. In order to make the changes you want to make, it is important to remember that culture change is both possible and achievable. By approaching it consciously and consistently, we can cultivate culture where people feel valued, respected and empowered to contribute their best. As leaders and change agents, I encouraged you to embrace this challenge with optimism, determination and discipline. Be the difference that makes the difference.

As we look ahead to the workplaces of the future, one thing becomes abundantly clear: taking a conscious approach to culture development will be more essential than ever. The evolving nature of workplace environments, coupled with shifting cultural dynamics, necessitates a proactive approach to step out of the culture you have to the culture you need.

Driven by growing awareness and societal shift, I believe that mental health in the workplace will continue to be a key focus area for organisations. The fallout from the COVID-19 pandemic has highlighted the need for proactive mental health initiatives, as employees continue to navigate new challenges such as remote work, social isolation and economic uncertainty. There is a growing need for workplace initiatives that destigmatise mental illness, as well as providing access to counselling and therapy services and promoting a culture of psychological safety and resilience. Organisations will need to develop wellbeing-oriented cultures, such as making mental health resources more accessible and better integrated into their system through employee assistance programmes and wellbeing platforms, if they want to meet this growing need.

Moreover, the increasing emphasis on diversity, equality and inclusion in society at large is reshaping expectations around workplace culture. Employees are now demanding workplace environments where they feel valued, cared-for, engaged and where they are able to 'bring their authentic selves' to work. Organisations that prioritise diversity and actively work to create an inclusive culture will be better positioned to attract and retain top talent in the future.

Finally, in the digital age, where remote work and virtual collaboration are becoming the norm, the traditional boundaries of the workplace are expanding. As a result, organisations must adapt to new modes of operation and communication, which will profoundly impact workplace culture. Cultivating a culture that values flexibility, adaptability and inclusivity will be crucial in fostering engagement and cohesion among remote and distributed teams. The rapid pace of technological advancement and forward march of globalisation are keeping the business landscape firmly in the Volatile, Uncertain, Complex and Ambiguous age. In such dynamic environments, a strong and resilient culture can serve as a stabilising force, providing employees with a sense of purpose, belonging and direction amid constant change.

As we navigate these evolving trends and challenges, it is clear to me that culture will play a key role in shaping the future success of organisations. Looking forward, I believe that active culture cultivation will be indispensable in creating successful workplaces that are adaptive, inclusive and resilient. By recognising the importance of culture and taking intentional steps to shape it, organisations can better position themselves for long-term sustainability and consistency in an age of complexity.

Index

Printed in the United States
by Baker & Taylor Publisher Services